Lecture Notes
in Control and Information Sciences 206

Editor: M. Thoma

Springer-Verlag London Ltd.

Zijad Aganović and Zoran Gajić

Linear Optimal Control of Bilinear Systems

with Applications to Singular Perturbations and Weak Coupling

Springer

Series Advisory Board

Authors

Zijad Aganović, President
Cylex Systems Inc., 6001 Broken Sound Parkway, Boca Raton, Florida 33487, USA

Zoran Gajić, Associate Professor
Department of Electrical and Computer Engineering, Rutgers University,
Piscataway, New Jersey 08855-0909, USA

ISBN 978-3-540-19976-2   ISBN 978-3-540-39378-8 (eBook)
DOI 10.1007/978-3-540-39378-8

British Library Cataloguing in Publication Data
A catalogue record for this book is available from the British Library

© Springer-Verlag London 1995
Originally published by Springer-Verlag London Limited in 1995

Typesetting: Camera ready by authors

69/3830-543210 Printed on acid-free paper

To Selma, Jaca, Adi & Veki

Preface

This book presents linear methods for optimal control of bilinear systems with emphasis on singularly perturbed and weakly coupled bilinear systems. The unified theme of this book is the use of reduced-order subproblems to simplify and decompose computations required for the optimal control of bilinear systems. There are numerous examples of bilinear control systems that provide great challenges to engineers, mathematicians, and computer scientists. Some of the examples of bilinear control systems are nuclear reactors, thermal processes, basic law of mass action, dynamics of heat exchanger with controlled flow, distillation columns, some processes in elasticity, dc motor, induction motor drives, mechanical brake system, aerial combat between two aircrafts and missile intercept problem. Many biological processes possess bilinear structures such as water balance and temperature regulation in human body, control of carbon dioxide in lungs, blood pressure, immune system, cardiac regulator, circulation of thyroxin in human body, respiratory chemostat, behavior of sense organ, hormone regulation, and kidney water balance. In addition, some economic processes (a growth of a national economy), processes in ecology and socioeconomics can be studied by the use of bilinear models.

The book studies the special classes of bilinear-quadratic control problems, namely, the singularly perturbed and weakly coupled bilinear control systems. In these cases, the obtained results are further simplified by producing the linear controllers at subsystem levels (slow and fast subsystems for singular perturbations) whose compositions produce the optimal linear global controllers. To demonstrate usefulness of the presented methods for singularly perturbed and weakly coupled bilinear systems, and to point out its various advantages, we have included several real control system examples such as: control of induction motor drives, chemical reactor, and paper making machine.

This book is intended for a wide readership, including control engineers, applied mathematicians, computer scientists, and advanced graduate students who seek a comprehensive view of current developments in the theory of optimal control of bilinear systems. The book emphasizes mathematical developments as well as their applications to solving practical problems without requiring a strong mathematical background.

The authors hope that this book will reduce some of the barriers that exist in recognizing the power and usefulness of the linear approach to optimal control of bilinear systems, and that it will help to broaden their implementation in practice. Also, we hope that this book will motivate some researchers to develop the corresponding algorithms for other classes of bilinear control systems by using the dynamic programming and its successive approximations method, and hopefully to extend these results to nonlinear-quadratic optimal control problems.

The authors are thankful for support and contributions from Professors M. Lim, B. Petrovic, X. Shen, W. Su, and graduate student J. Rutkowski. We also appreciate helpful discussions with Professors E. Tzanakou and E. Sontag from Rutgers University, and Dr. R. Srikant from AT&T Bell Laboratories.

Authors
Piscataway, NJ, USA
April 1995

Contents

Chapter 1

Introduction

In between of linear and nonlinear systems lies a very large class of so called bilinear systems. They represent an enormous number of the real world phenomena (Mohler, 1970, 1973, 1974, 1991; Mohler and Chen, 1970; Mohler and Kolodziej, 1980; Bruni et al., 1974; Bahrami and Kim, 1975; Sundareshan and Fundkowski, 1985, 1986; Williamson, 1977; Espana and Landau, 1978). The bilinear control systems are described by the following evolution equation

$$\dot{x} = \left(A + \sum_{k=1}^{m} N_k u_k \right) x + Bu \qquad (1.1)$$

where $x \in \Re^n$ is the state vector and $u \in \Re^m$ is the control vector. The matrices A, B, and $N_k \in \Re^{n \times n}$, $k = 1, ..., m$ are of appropriate dimensions. The product of state and control variables, that is $u_k x$, distinguishes these classes of systems from the linear ones, but at the same time makes bilinear systems so general such that "every input-output map $\mathcal{F}$ can be approximated as closely as desired by maps which arise from bilinear systems, provided that $\mathcal{F}$ satisfies certain continuity and causality conditions" (Sussmann, 1976; Lo, 1975). That is why A. Balakrishnan raised an important question: "Are all nonlinear systems bilinear," (Balakrishnan, 1976).

Very often in the literature on the bilinear control systems, the mathematical model (1.1) is recorded as

$$\dot{x} = Ax + \left(B + \sum_{j=1}^{n} M_j x_j \right) u \qquad (1.2)$$

with $M_j \in \Re^{n \times m}$.

A general block diagram for a bilinear control system, represented by (1.1) is given in Figure 1.1.

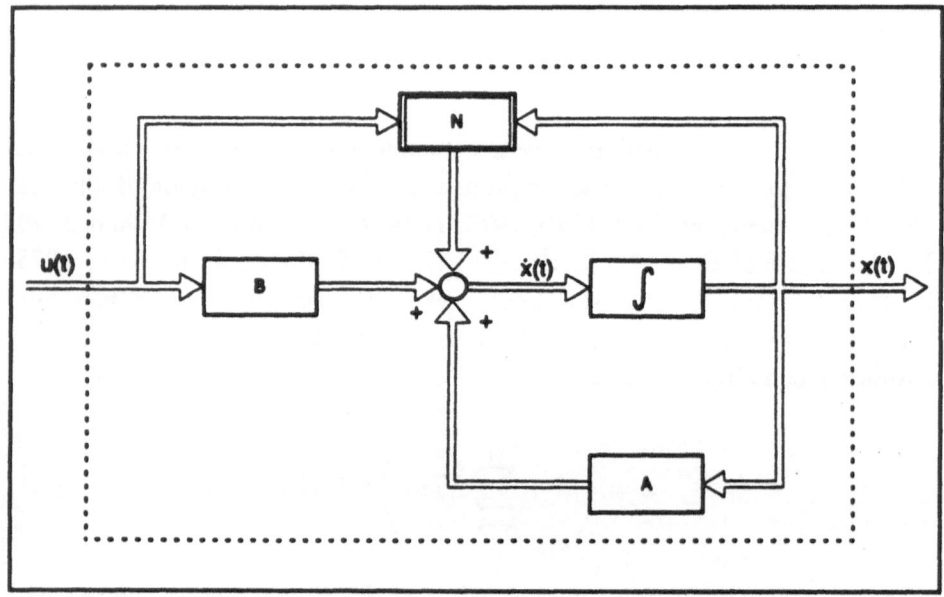

Figure 1.1: Block diagram for a general bilinear control system

The importance of bilinear systems has been recognized since the work of (Wiener, 1948), who believed that they are the essence of understanding the behavior of neural and biological computing networks. Bilinear systems were brought to attention of control engineers due to work of R. Mohler in the early seventies. Originally they were introduced in the study of nuclear reactors (Mohler, 1970, 1973; Mohler and Chen, 1970; Mohler and Kolodziej, 1980), where the bilinearity of state and control variables appears naturally in the reactor kinetic equation. Related problems such as nuclear fission, reactor shut down, reactor control and thermal processes are described by bilinear dynamic equations also. During the seventies and the first part of the eighties the bilinear control systems were the subject of very extensive research. In the last decade they become an integral part of the modern nonlinear trend in control and system theory so that no too many research studies dealt strictly with the bilinear systems. However, many important results obtained for the nonlinear control systems can be specialized and used for the bilinear control systems.

The major importance of bilinear systems indeed lies in their applications to the real world systems as demonstrated in the following paragraphs. Bilinear systems naturally represent many physical processes, for example: the basic law of mass action (Mohler, 1970, Bruni et al., 1974), dynamics of heat exchanger with controlled flow (Bruni et al., 1974); distillation columns (Espana and Landau, 1978); some processes in elasticity (Slemrod, 1978); dc motor (Bruni et al., 1974); induction motor drives (Figalli et al., 1984); mechanical brake system (Mohler, 1970, 1973); aerial combat between two aircrafts and missile intercept problem (Wei and Pearson, 1978); modeling and control of a small furnace, (Baheti and Mohler, 1981); control of hydraulic rotary multi-motor systems, (Guo et al., 1994).

Many biological processes, such as the population dynamics of biological species (Mohler, 1970; Bruni et al., 1974); water balance and temperature regulation in human body (Mohler, 1970, 1973, 1974); control of carbon dioxide in lungs (Mohler, 1970); blood pressure (Mohler, 1991); immune system (Mohler, 1991); cardiac regulator (Mohler, 1970); circulation of thyroxin in human body (Mohler, 1970); behavior of sense organ (Bruni et al., 1974); biological control (Williamson, 1977); growth of cancer cell population and finding the optimal therapy (Bahrami and Kim, 1975; Sundareshan and Fundkowski, 1985, 1986); respiratory chemostat, hormone regulation, kidney water balance, (Mohler, 1991); dissolved oxygen process (Ko et al., 1982); cell mass concentration in continuous culture (Yi et al., 1989) — all of them are described by bilinear models.

Some economic processes (a growth of a national economy (Bruni et al., 1974), processes in ecology and socioeconomics (Mohler, 1973) may be studied by the use of bilinear models. The bilinear systems are the adaptive ones (Mohler, 1970). They have the variable structure strongly dependent on the control vector (1.1). Thus, the study of bilinear systems might bring some interesting results in the field of adaptive systems as well (Ionesku and Monopoli, 1975).

Despite vast applications of bilinear systems, they have not been studied extensively in the domain of singular perturbations and weak coupling except for a few minor results (Guillen and Armada, 1980; Tzafestas and Anagnostou, 1984a, 1984b; Asamoah and Jamshidi, 1987). In this book we will pay special attention to the singularly perturbed and weakly coupled bilinear control systems and derive techniques for decompositions of these systems into subsystems. This simplifies implementation of the control algorithms, speeds up real-time control and signal processing (filtering), and introduces parallelism in the design procedures. The above features will be facilitated by exploiting the presence of a small perturbation parameter, which in the case of singularly perturbed systems introduces numerical ill-conditioning. However, having obtained the system decomposition into subsystems (corresponding to the slow and fast time scales) the numerical ill-conditioning is removed.

1.2 Singularly Perturbed and Weakly Coupled Bilinear Control Systems

The application of bilinear systems is very well documented. A little bit is known about the fact that a large number of these systems display the multi time scale property or the singularly perturbed structure. In this section we present some real world bilinear singularly perturbed and weakly coupled control systems.

The theory of singular perturbations has been a rapidly developing and highly recognized research area of control engineering in the last twenty five years. Almost all important control aspects for linear systems have been studied so far and valuable and practically implementable results have been obtained. The extension of these results to nonlinear systems happened to be a difficult task. Only under very restrictive conditions and for very limited classes of nonlinear systems some results were obtained (O'Malley, 1974;

Chow and Kokotovic, 1978a, 1978b, 1981; Suzuki, 1981; Saberi and Khalil, 1984, 1985).

The singularly perturbed bilinear control system consistent with (1.1) is described by the following differential equation

$$
\begin{bmatrix} \dot{y} \\ \epsilon\dot{z} \end{bmatrix} = \begin{bmatrix} A_1 & A_2 \\ A_3 & A_4 \end{bmatrix} \begin{bmatrix} y \\ z \end{bmatrix} + \begin{bmatrix} B_1 \\ B_2 \end{bmatrix} u + \sum_{k=1}^{m} u_k \begin{bmatrix} N_1^k & N_2^k \\ N_3^k & N_4^k \end{bmatrix} \begin{bmatrix} y \\ z \end{bmatrix} \tag{1.3}
$$

where $y \in \Re^{n_1}$ is the slow state vector, $z \in \Re^{n_2}$ is the fast state vector, $u \in \Re^m$ is the control input and ϵ is a small positive parameter. Constant matrices A_i, N_i^k; $i = 1, 2, 3, 4$, and B_1 and B_2 are of appropriate dimensions. In the following some important bilinear systems in biological and physical sciences that display singularly perturbed structure will be introduced.

The neutron level control problem in a fission reactor (Mohler, 1970, 1973; Mohler and Chen, 1970) is described by the following equation

$$
\begin{bmatrix} \dot{n} \\ \dot{c} \end{bmatrix} = \begin{bmatrix} -\frac{\beta}{\lambda} & \alpha \\ -\frac{\beta}{\lambda} & -\alpha \end{bmatrix} \begin{bmatrix} n \\ c \end{bmatrix} + u \begin{bmatrix} \frac{1}{\lambda} & 0 \\ 0 & 0 \end{bmatrix} \begin{bmatrix} n \\ c \end{bmatrix} \tag{1.4}
$$

where n is neutron population, c is precursor population, u is reactivity (it represents a control input), and α, β, and λ are known constants. It is important to point out that λ takes very small values. Typical values for these constants clearly show the singularly perturbed structure of (1.4), (Quin, 1980), $\lambda = 0.00001$, $\beta = 0.0065$, and $\alpha = 0.4$. Introducing a transformation $x_1 = \lambda n$ and $x_2 = c$ one gets

$$
\begin{bmatrix} \lambda \dot{x}_1 \\ \dot{x}_2 \end{bmatrix} = \begin{bmatrix} -\beta & \alpha \\ \beta & -\alpha \end{bmatrix} \begin{bmatrix} x_1 \\ x_2 \end{bmatrix} + u \begin{bmatrix} 1 & 0 \\ 0 & 0 \end{bmatrix} \begin{bmatrix} x_1 \\ x_2 \end{bmatrix} \tag{1.5}
$$

that is, a singularly perturbed bilinear form.

The mathematical model of a dc-motor (Bruni et al., 1974) has a bilinear form

$$
\begin{bmatrix} \dot{x}_1 \\ \dot{x}_2 \end{bmatrix} = \begin{bmatrix} -\frac{R}{L} & 0 \\ 0 & -\frac{F}{J} \end{bmatrix} \begin{bmatrix} x_1 \\ x_2 \end{bmatrix} + u_1 \begin{bmatrix} \frac{1}{L} \\ 0 \end{bmatrix} + u_2 \begin{bmatrix} 0 & K \\ \frac{K}{J} & 0 \end{bmatrix} \begin{bmatrix} x_1 \\ x_2 \end{bmatrix} \tag{1.6}
$$

where x_1 and x_2 are the rotor current and the axis speed (state variables), u_1 and u_2 are the stator current and the rotor voltage (control variables), R

and L are electric parameters of rotor. F and J are mechanical parameters of the load and K is the torque constant. Since $L \ll J$ is a well known fact, the singularly perturbed structure of (1.6) is obvious.

The singularly perturbed bilinear control system consistent with form (1.2) is represented by

$$\begin{bmatrix} \dot{y} \\ \epsilon\dot{z} \end{bmatrix} = \begin{bmatrix} A_1 & A_2 \\ A_3 & A_4 \end{bmatrix} \begin{bmatrix} y \\ z \end{bmatrix} + \begin{bmatrix} B_1 \\ B_2 \end{bmatrix} u + \left\{ \begin{bmatrix} y \\ z \end{bmatrix} \begin{bmatrix} M_s \\ M_f \end{bmatrix} \right\} u \qquad (1.7)$$

with initial conditions

$$\begin{bmatrix} y(t_0) \\ z(t_0) \end{bmatrix} = \begin{bmatrix} y^0 \\ z^0 \end{bmatrix}$$

where $y \in \Re^{n_1}$, $z \in \Re^{n_2}$ are, respectively, slow and fast state variables, ϵ is a small positive parameter, and

$$\left\{ \begin{bmatrix} y \\ z \end{bmatrix} \begin{bmatrix} M_s \\ M_s \end{bmatrix} \right\} = \sum_{j=1}^{n_1} y_j \begin{bmatrix} M_{sj} \\ M_{fj} \end{bmatrix} + \sum_{j=n_1+1}^{n_1+n_2} z_j \begin{bmatrix} M_{sj} \\ M_{fj} \end{bmatrix} \qquad (1.8)$$

The following notation is used in order to relate (1.2) and (1.7)

$$A = \begin{bmatrix} A_1 & A_2 \\ \frac{A_3}{\epsilon} & \frac{A_4}{\epsilon} \end{bmatrix}, \quad B = \begin{bmatrix} B_1 \\ \frac{B_2}{\epsilon} \end{bmatrix}, \quad M = \begin{bmatrix} M_s \\ \frac{M_f}{\epsilon} \end{bmatrix}, \quad x(t) = \begin{bmatrix} y(t) \\ z(t) \end{bmatrix} \qquad (1.9)$$

The bilinear model of induction motor drives is given by a fourth-order differential equation (Figali et al., 1984). This frequency controlled two phase induction motor can be put in the singularly perturbed form (1.7) as given below. The state and control variable are

$$x = \begin{bmatrix} y_1 \\ y_2 \\ z_1 \\ z_2 \end{bmatrix} = \begin{bmatrix} \phi_{ds} \\ \phi_{qs} \\ i_{ds} \\ i_{qs} \end{bmatrix}, \quad u = \begin{bmatrix} u_1 \\ u_2 \\ u_3 \end{bmatrix} = \begin{bmatrix} v_{ds} \\ v_{qs} \\ \omega_s \end{bmatrix}$$

where

ϕ_{ds} and ϕ_{qs} — projections of the stator flux
i_{ds} and i_{qs} — projections of the stator current
v_{ds} and v_{qs} — projections of the supply voltage
ω_s — slip angular frequency.

The problem matrices have the following values

$$
A = \begin{bmatrix} 0 & 321.57 & -0.312 & 0 \\ -312.57 & 0 & 0 & -0.312 \\ 98.87 & 27059 & -44.93 & 2.57 \\ -27059 & 98.87 & -2.57 & -44.93 \end{bmatrix}, \quad x(t_0) = \begin{bmatrix} -0.07 \\ 0.04 \\ 15 \\ 47 \end{bmatrix}
$$

$$
B = \begin{bmatrix} 1 & 0 & 0 \\ 0 & 1 & -7.3 \\ 87.3 & 0 & 87.8 \\ 0 & 87.3 & -53 \end{bmatrix}, \quad M_1 = \begin{bmatrix} 0 & 0 & 0 \\ 0 & 0 & -1 \\ 0 & 0 & 0 \\ 0 & 0 & 0 \end{bmatrix}
$$

$$
M_2 = \begin{bmatrix} 0 & 0 & 1 \\ 0 & 0 & 0 \\ 0 & 0 & 0 \\ 0 & 0 & 0 \end{bmatrix}, \quad M_3 = \begin{bmatrix} 0 & 0 & 0 \\ 0 & 0 & 0 \\ 0 & 0 & 0 \\ 0 & 1 & 0 \end{bmatrix}, \quad M_4 = \begin{bmatrix} 0 & 0 & 0 \\ 0 & 0 & 0 \\ 0 & 0 & 1 \\ 0 & 0 & 0 \end{bmatrix}
$$

It can be easily seen (big entries in the last two rows) that z_1 and z_2 are fast variables and y_1 and y_2 are slow variables; that is, this system displays two time scale property also.

Many other real world biological systems either have or can be brought in the singularly perturbed form. For example, regulation of carbon dioxide in the respiratory system (Mohler, 1970, 1991), where time constants corresponding to two time scales are determined by the lung and tissue reservoir volumes, respectively denoted by V_1 and V_2. The corresponding mathematical model is given by

$$
\dot{x}_1 = \frac{1}{V_1}[u(c_1 - x_1) + c_2(x_2 - c_3 x_1 - c_4)]
$$

$$
(1.10)
$$

$$
\dot{x}_2 = \frac{1}{V_2}[c_5 - c_2(x_2 - c_3 x_1 - c_4)]
$$

where c_i, $i = 1, ..., 5$, are known parameters. The state variables x_1 and x_2 respectively represent the rates of change of the lung and tissue concentration of CO_2.

The mechanical portion of the cardiovascular system is described by the following singularly perturbed bilinear control system (Mohler, 1991)

$$
\begin{aligned}
\lambda_1 \dot{x}_1 &= u_1(x_2 - x_1) + v_1 \\
\lambda_2 \dot{x}_2 &= u_1(x_1 - x_2) - v_2 \\
\lambda_3 \dot{x}_3 &= u_2(x_2 - x_3) + v_2 \\
\lambda_4 \dot{x}_4 &= u_2(x_3 - x_4) - v_1
\end{aligned}
\tag{1.11}
$$

where x_1, x_2, x_3, and x_4 are compartmental pressures in the arteries filled from the left heart, veins to the right heart, arteries from the right heart, and veins to the left heart, respectively; v_1 and v_2 are cardiac outputs from the left and right heart, respectively; $\lambda_1, \lambda_2, \lambda_3$, and λ_4 are corresponding compartmental capacities (assumed to be constant).

In addition, the mechanical brake system (Mohler, 1970), which is in fact a system of bilinear differential equations with a huge parameter (car mass) multiplying some derivatives (Desoer and Shena, 1970), can be put in the singularly perturbed bilinear form. The distillation columns (Espana and Landau, 1978) are described by three time scale bilinear models involving huge parameters also. In (Cronin, 1987), it has been shown that the singular perturbation theory is the most efficient tool in the study of the famous Hodgkin-Huxley model of nerve conduction. Due to very complex and non-linear structure of this equation it is hard to believe that its linearized model will produced satisfactory results. However, the bilinearization procedure will considerably improve the approximation and it might result in a better understanding of the neural conductivity.

The weakly coupled bilinear control system, in the representation consistent with form (1.2), is represented by

$$
\begin{bmatrix} \dot{y}_1 \\ \dot{y}_2 \end{bmatrix} = \begin{bmatrix} A_1 & \epsilon A_2 \\ \epsilon A_3 & A_4 \end{bmatrix} \begin{bmatrix} y_1 \\ y_2 \end{bmatrix} + \begin{bmatrix} B_1 & \epsilon B_2 \\ \epsilon B_3 & B_4 \end{bmatrix} \begin{bmatrix} u_1 \\ u_2 \end{bmatrix}
$$

$$
+ \left\{ \begin{bmatrix} y_1 \\ y_2 \end{bmatrix} \begin{bmatrix} M_a & \epsilon M_b \\ \epsilon M_c & M_d \end{bmatrix} \right\} \begin{bmatrix} u_1 \\ u_2 \end{bmatrix}, \quad \begin{bmatrix} y_1(t_0) \\ y_2(t_0) \end{bmatrix} = \begin{bmatrix} y_1^0 \\ y_2^0 \end{bmatrix}
\tag{1.12}
$$

where $y_1 \in \Re^{n_1}$, $y_2 \in \Re^{n_2}$, $u_i \in \Re^{m_i}$, $i = 1, 2$, and ϵ is a small coupling parameter, with the following notation

$$\left\{ \begin{bmatrix} y_1 \\ y_2 \end{bmatrix} \begin{bmatrix} M_1 & \epsilon M_2 \\ \epsilon M_3 & M_4 \end{bmatrix} \right\} = \sum_{i=1}^{n_1} y_{1i} \begin{bmatrix} M_{ai} & M_{bi} \\ M_{ci} & M_{di} \end{bmatrix}$$

(1.13)

$$+ \sum_{j=n_1+1}^{n_1+n_2} y_{2(j-n_1)} \begin{bmatrix} M_{aj} & M_{bj} \\ M_{cj} & M_{dj} \end{bmatrix}$$

where $M_{ai} \in \Re^{n_1 \times m_1}$, $M_{bi} \in \Re^{n_1 \times m_2}$, $M_{ci} \in \Re^{n_2 \times m_1}$, $M_{di} \in \Re^{n_2 \times m_2}$.

The weakly coupled bilinear control systems are either naturally weakly coupled or they can be obtained in the process of bilinearization of nonlinear weakly coupled control systems. The natural bilinear weakly coupled control system is for example the problem of a paper making machine as given in (Ying et al., 1992). The bilinear mathematical model of this system is formulated according to (1.12) and (1.13) as

$$A = \begin{bmatrix} -1.93 & 0 & 0 & 0 \\ 0.394 & -0.426 & 0 & 0 \\ 0 & 0 & -0.63 & 0 \\ 0.095 & -0.103 & 0.413 & -0.426 \end{bmatrix}, \quad B = \begin{bmatrix} 1.274 & 1.274 \\ 0 & 0 \\ 1.34 & -0.65 \\ 0 & 0 \end{bmatrix}$$

(1.14)

$$M_1 = \begin{bmatrix} 0 & 0 \\ 0 & 0 \\ 0.755 & 0.366 \\ 0 & 0 \end{bmatrix}, \quad M_2 = \begin{bmatrix} 0 & 0 \\ 0 & 0 \\ 0 & 0 \\ 0 & 0 \end{bmatrix}$$

$$M_3 = \begin{bmatrix} 0 & 0 \\ 0 & 0 \\ -0.718 & -0.718 \\ 0 & 0 \end{bmatrix}, \quad M_4 = \begin{bmatrix} 0 & 0 \\ 0 & 0 \\ 0 & 0 \\ 0 & 0 \end{bmatrix}$$

Note that for this model only the matrix A is weakly coupled, whereas the matrices B, M_1, and M_2 have no weakly coupled forms. Since in this book we developed the linear approach to almost all bilinear-quadratic control problems, we can use the results from (Skataric et al., 1991), in which, it has been shown that the classes of linear-quadratic optimal control problems having weakly coupled system matrix and strongly coupled input matrix (quasi weakly coupled systems) can be studied as the weakly coupled linear-quadratic optimal control problems by assuming the special form for the state penalty matrix. This can be achieved, in this particular example, by using the following weighting matrices Q and R

$$
Q = \begin{bmatrix} 1 & 0 & 0.13 & 0 \\ 0 & 1 & 0 & 0.09 \\ 0.13 & 0 & 0.1 & 0 \\ 0 & 0.09 & 0 & 0.2 \end{bmatrix} = \begin{bmatrix} O(1) & O(\epsilon) \\ O(\epsilon) & O(\epsilon) \end{bmatrix}, \quad R = \begin{bmatrix} 1 & 0 \\ 0 & 1 \end{bmatrix} \quad (1.15)
$$

Small perturbation parameter is $\epsilon = 0.1$.

Many other nonlinear singularly perturbed systems can be brought into the singularly perturbed bilinear form by performing bilinearization (Schwartz, 1988).

1.2 Book Organization

This book consists of five chapters. Chapter 1 comprises an introduction on the general bilinear control systems and presents several examples of the real world singularly perturbed and weakly coupled bilinear control systems. After the introductory chapter, in the first part of this book, in Chapters 2, and 3, we study the linear optimal control of singularly perturbed and weakly coupled bilinear control systems. In Chapter 4 we consider new techniques for optimization of bilinear control systems, and in Chapter 5 some future research directions are outlined.

In Chapter 2 we study the optimization of singularly perturbed bilinear control systems. A sequence of linear state and costate equations is constructed, and the open-loop solution of the optimization problem is obtained in terms of the reduced-order slow and fast subsystems. The ill-defined numerical problem is completely decomposed into slow and fast time scales, leading to the reduction in the size of the required computations and allowing parallel processing of information. In addition, the near-optimal "closed-loop" control is obtained in the form of a linear approximate "feedback" control law as a linear combination of the slow and fast variables (the composite control), with the coefficient matrices calculated from two reduced-order independent time varying linear-quadratic optimal control problems. The composite control is an $O(\epsilon)$ close to the optimal one, where ϵ stands for a small singular perturbation parameter. The composite control implies an $O(\epsilon)$ closeness of the system trajectories to the optimal ones, and an $O(\epsilon)$ approximation for the performance criterion. An algorithm that produces an arbitrary order of accuracy for the closed-loop approximate "feedback" control is also presented in this chapter. As a case study we present the approximate linear controller design of a bilinear model of induction motor drives (Figalli et al., 1984).

In Chapter 2 and 3 we use the classic approach to continuous-time deterministic singularly perturbed and weakly coupled bilinear systems following the results of (Hofer and Tibken, 1984) and (Cebuhar and Constanza, 1988). In both cases decompositions of the original systems are achieved so that the results are obtained in terms of the reduced-order subsystems. Chapter 3 considers the optimization of the time invariant bilinear weakly coupled system with a quadratic performance criterion. It has been shown how to design the linear laws by using the linear local controllers whose composition produce the optimal linear global controller. The corresponding results have been obtained for both the open-loop and approximate "closed-loop" control of bilinear weakly coupled systems. The presented methodology is demonstrated on a real physical system, a bilinear-quadratic optimal control problem of a paper making machine (Ying et al., 1992).

The presented work on singularly perturbed bilinear control systems is mostly based on the doctoral dissertation (Aganovic, 1993), and the one on weakly coupled bilinear systems on (Aganovic, 1993; Aganovic and Gajic, 1993).

In the second part of the book, in Chapter 4, the modern approach based on the successive approximation technique for optimal control of general bilinear-quadratic systems is presented. The obtained results follows closely the recent work by the authors (Aganovic, 1993; Aganovic and Gajic, 1994, 1995). Both the steady state and finite-time optimization problems are considered. The presented results produce the linear optimal control law and simplify the already existing results of (Hofer and Tibken, 1984) and (Cebuhar and Constanza, 1988). Namely, it has been shown how to replace the sequences of the Riccati equations by the sequences of the Lyapunov equations whose solutions comprise the approximate linear controller for bilinear systems. In the limit these controllers tend to the optimal ones. The proposed schemes are very efficient. It takes only a few iterations to get the convergence to the optimal controller gains. The main themes of the results obtained in Chapter 5 are the linearity of the optimal control law and the power of dynamic programming and its successive approximations technique used to get the desired results.

In Chapter 5 we indicate some future research problems related with the singularly perturbed and weakly coupled bilinear systems. We propose the extension of the results presented in Chapters 2 and 3 in the spirit of results obtained in Chapter 4. Also, we indicate importance of studying singularly perturbed and weakly coupled bilinear systems in the discrete-time domain. In addition, we indicate potential research on singularly perturbed and weakly coupled bilinear stochastic systems, and related linear systems with state-dependent and/or control-dependent disturbances that are described by the bilinear stochastic modes.

Chapter 2

Continuous-Time Singularly Perturbed Bilinear Systems

2.1 Introduction

The theory of singular perturbations has been a highly recognized and rapidly developing area of control systems research in the last twenty five years (Kokotovic et al., 1986; Kokotovic and Khalil, 1986; Gajic et al., 1990). Almost all important control aspects for linear systems have been studied so far and valuable and practically implementable results have been obtained. The extension of these results to the nonlinear systems happened to be a very difficult task. Only under very restrictive conditions and for very limited classes of nonlinear systems some results were obtained (Saberi and Khalil, 1984, 1985; O'Malley, 1974a, 1974b; Chow and Kokotovic, 1978a, 1978b, 1981; Suzuki, 1981).

In between of linear and nonlinear systems lies a very large class of so-called bilinear systems (Mohler, 1991). This class of "nearly linear" systems has not been studied so far in the context of singular perturbations, except for a few minor attempts (Guillen and Armada, 1980; Tzafestas and Anagnostou, 1984a; Asamoah and Jamshidi, 1987).

In the previous chapter we have indicated that a large number of real world systems are bilinear. In addition, many real physical systems possess the structure of the singularly perturbed bilinear control systems such as: neutron level control problem in a fission reactor (Mohler, 1973), dc-motor (Bruni et al., 1974), induction motor drives (Figalli et al., 1984), regulation of carbon-dioxide in the respiratory system (Mohler, 1970), mechanical brake system (Mohler, 1970), and distillation columns (Espana and Landau, 1978).

The purpose of this chapter is to study the optimal control problem of singularly perturbed bilinear systems with a quadratic performance criterion. We consider both the open-loop and "closed-loop" optimal control problems. An introductory review of existing results on optimization and stabilization of bilinear systems that are related to the material studied in this book is presented in Section 2.2.

In the first part of this chapter, we study the optimal open-loop control problem of singularly perturbed bilinear systems with a quadratic performance criterion. The obtained results utilize the recursive scheme for the optimal control of a general bilinear system with a quadratic performance criterion (Hofer and Tibken, 1988) and the time varying version of the reduced-order method with an arbitrary degree of accuracy for solving the linear-quadratic optimal open-loop singularly perturbed control problem (Su et al., 1992a). This problem is solved as a sequence of linear two-point boundary value singularly perturbed problems. At each iteration step the ill-conditioned linear time varying two-point boundary value problem is transformed in the pure-slow and pure-fast completely decoupled initial value problems. By doing this, the stiffness of the singularly perturbed two-point boundary value problem is converted in the problem of an ill-defined linear system of algebraic equations. However, the latter problem is much easier to handle. The size of required computations is reduced since the introduced transformation allows parallel processing of information.

In Section 2.4, we utilize the idea of the composite control law for singularly perturbed systems (Saberi and Khalil, 1985; Suzuki, 1981; Chow and Kokotovic, 1976), and the recursive scheme for the optimal control of a general bilinear system with a quadratic performance criterion (Cebuhar and Constanza, 1984). The obtained composite control law for singularly perturbed bilinear systems is represented by a linear combination of the slow and fast variables. The matrix coefficients for this linear combination are obtained from the recursive scheme applied to the two reduced-order independent time varying linear-quadratic control problems. The composite control law is $O(\epsilon)$ close to the optimal one, which implies the $O(\epsilon)$ closeness of the near-optimal trajectories to the optimal ones, and the $O(\epsilon)$ approximation for the performance criterion. A real world numerical example, an induction motor drives, is used to demonstrate the efficiency of the obtained composite control. In addition, an algorithm for achieving higher order approximations is proposed in the spirit of the recursive methods for singularly perturbed control systems developed by Gajic and his coworkers (Gajic et al., 1990; Gajic and Shen, 1993).

The results obtained in this chapter are mostly based on the doctoral dissertation (Aganovic, 1993).

2.2 An Overview of Existing Results

In this section we give a brief overview of the existing results on the optimal control and the related stabilization problems of bilinear systems. At the present time there are no general methods for optimal regulation of bilinear systems in spite of the fact that many papers on this topic have recently appeared in the literature, see for example (Gutman, 1981; Ryan, 1984; Banks and Yew, 1985, 1986; Tzafestas et al., 1984; Mohler, 1991). Several different approaches have been taken so far to solve the optimization problem of bilinear systems, see (Mohler and Kolodziej, 1980; Mohler, 1991) and references therein.

In (Slemrod, 1978; Quin, 1980; Ryan and Buckingham, 1983) the quadratic feedback controls are derived for special classes of bilinear systems with a purely imaginary spectrum of the system matrix A. The approach taken in (Slemrod, 1978) is very interesting from the engineering point of view. A stabilizing controller is chosen such that the derivative of a Lyapunov function along the system trajectories is negative semidefinite. It has been shown that such a control is the optimal one for the special classes of cost functionals that are quadratic in state and control variables. The stabilization and optimization technique of (Slemrod, 1978) is extended in (Figalli et al., 1984) to the design of an optimal feedback controller of induction motor drives. Quadratic polynomial controls that stabilize given bilinear system by using a hiperstability technique are derived in (Ionescu and Monopoli, 1975). The applicability of the quadratic control law to the case when the A matrix has an arbitrary spectrum is studied in (Gutman, 1981). The stability problem of time varying bilinear systems with output feedback is considered in (Chen et al., 1991).

In the papers (Longchamp, 1980a, 1980b) the well known techniques of linear systems are extended to the bilinear systems. The first paper considers a scalar input and the second one (Longchamp, 1980b) deals with a vector input and the stable system matrix. The problem of designing a linear time-invariant stabilizing feedback law in the form $u = Kx$ is considered in (Derese and Noldus, 1980). The feedback gain K and the corresponding stability region are obtained in terms of the very well-known algebraic Riccati equation so that the linear optimal control theory is extended to bilinear control systems.

For the general regulation problem of bilinear systems, with the exception of the simplest cases, it is not possible to express the optimal control in the explicit feedback form. The minimum energy optimal control for the commutative bilinear systems $(AN_k = N_kA, \ k = 1, 2, ..., m)$ is studied in (Wei and Pearson, 1978; Banks and Yew 1986). It is shown that in such a case the optimal control is given by a constant. In the general, noncommutative case, the optimal control for a single input is obtained by considering the Lie algebra generated by the system matrices (Banks and Yew, 1986). A single input discrete system with a cost functional quadratic in state only is considered in (Swamy and Tran, 1979).

A very interesting optimization problem is presented in (Biran and McInnus, 1979), where the optimal control theory is applied in the estimation of the time varying effects of antitumor drugs on the kinetics of the cell cycle.

In order to get the feedback solution in a simple form a cost functional was modified in (Ryan, 1984) by the inclusion of an additional nonnegative state function. This function "regulates" the problem, but it makes it quite artificial. In (Banks and Yew, 1985) the tensor theory in the Hilbert spaces is used in order to define a suboptimal control in terms of truncated power series.

The optimal control results of bilinear systems as derived in (Tzafestas et al., 1984) resemble to the linear-quadratic optimal control theory. Namely, stabilization and optimization problems of bilinear systems are solved in terms of the Lyapunov and Riccati equations. For a single input bilinear system having the form (1.1) and the following performance criterion

$$J = \frac{1}{2} \int\limits_{t_0}^{t_f} \left[x^T(t)K(x,t)x(t) + R(t)u^2(t) \right] dt + \frac{1}{2} x^T(t_f) P_f x(t_f) \quad (2.1)$$

the feedback control law is obtained as

$$u(x) = -R^{-1}(t)S^T(x)P(t)x \quad (2.2)$$

where S, P, and K are given by

$$S(x) = B + Nx$$
$$-\dot{P} = A^T P + PA - PBR^{-1}B^T P + Q, \quad P(t_f) = P_f \quad (2.3)$$
$$K(x,t) = Q(t) + P(t)\left[S(x)R^{-1}S^T(x) - BR^{-1}B^T\right]P(t)$$

It is shown that the optimal control (2.2) is stabilizing. It can be seen that this control contains two terms the linear feedback $-R^{-1}B^T Px$ and the quadratic feedback $R^{-1}x^T N^T Px$. The corresponding results for a vector input are also derived in (Tzafestas et al., 1984).

In (Ying et al., 1993) a suboptimal control law for a general bilinear system is derived as a sum of two components: the optimal linear-quadratic control law component and the correction term quadratic in state that compensates for the present of the bilinear term in the state equation.

The recursive schemes for near-optimal control of bilinear systems with quadratic performance criterion have been obtained in (Cebuhar and Constanza, 1984) and (Hofer and Tibken, 1988). Because of their importance for the material presented in Chapters 2 and 3 the main results of these papers will be presented later.

2.3 Open-Loop Optimal Control of Singularly Perturbed Bilinear Systems

Consider the finite time optimal control problem of a bilinear system represented in a symbolic notation by

$$\dot{x} = Ax + Bu + \{xM\}u, \qquad x(0) = x_0 \qquad (2.4)$$

with a quadratic performance criterion to be minimized

$$J = \frac{1}{2} \int_{t_0}^{t_f} (x^T Q x + u^T R u) \, dt \qquad (2.5)$$

In the above expressions $x \in \Re^n$ are state variables, $u \in \Re^m$ is a control input, A, B, M, Q, and R are constant matrices of appropriate dimensions, with $R = R^T > 0$, and $Q = Q^T \geq 0$. The notation used for the bilinear term in (2.4) means

$$\{xM\} = \sum_{j=1}^{n} x_j M_j, \qquad M_j \in \Re^{n \times m} \qquad (2.6)$$

From the Hamiltonian corresponding (2.4)-(2.5) and given by

$$H(x, u, \rho) = \frac{1}{2}(x^T Q x + u^T R u) + \rho^T (Ax + Bu + \{xM\}u) \qquad (2.7)$$

we get the expression for the finite time optimal open-loop control of a bilinear system as

$$u^* = -R^{-1}(B + \{xM\})^T \rho(t) \qquad (2.8)$$

where $\rho(t) \in \Re^n$ stands for the costate variables. The costate variable can be obtained from the following system of equations that was originally derived in (Hofer and Tibken, 1988)

$$\dot{x}_i = [Ax]_i - \left[(B + \{xM\})R^{-1}(B + \{xM\})^T \rho\right]_i, \qquad x_i(t_0) = x_i^0,$$

$$
\begin{aligned}
\dot{\rho}_i &= -[Qx]_i - [A^T \rho]_i \\
&+ \frac{1}{2}\rho^T \left[M_i R^{-1}(B + \{xM\})^T + (B + \{xM\})R^{-1}M_i^T\right]\rho, \\
&\rho_i(t_f) = [Fx(t_f)]_i
\end{aligned}
\tag{2.9}
$$

where $[...]_i$, $i = 1, ... , n$, is the i-th component of the corresponding vector. This two-point boundary value problem of the coupled nonlinear differential equations is not easy solvable. It is shown in (Hofer and Tibken, 1988) that the system (2.9) can be rewritten in the compact matrix form of the state-costate equations resembling to those of a linear-quadratic optimal control problem

$$\dot{x} = \tilde{A}x - \tilde{B}R^{-1}\tilde{B}^T\rho, \qquad x(t_0) = x^0,$$

$$\dot{\rho} = -\tilde{Q}x - \tilde{A}^T\rho, \qquad \rho(t_f) = Fx(t_f)$$

(2.10)

where $\tilde{A}$, $\tilde{Q}$, $\tilde{B}R^{-1}\tilde{B}^T$ are time varying matrices. Note that these matrices are functions of $x(t)$ and $\rho(t)$ so that the right-hand sides of (2.10) are nonlinear. The following linear two-point boundary value scheme has been proposed for solving (2.10), (Hofer and Tibken, 1988)

$$\dot{x}^{(k+1)} = \tilde{A}^{(k)}x^{(k+1)} - \tilde{B}^{(k)}R^{-1}\tilde{B}^{(k)^T}\rho^{(k+1)}$$

(2.11)

$$\dot{\rho}^{(k+1)} = -\tilde{Q}^{(k)}x^{(k+1)} - \tilde{A}^{(k)^T}\rho^{(k+1)}$$

with boundary conditions expressed in the standard form as

$$U\begin{bmatrix} x^{(k+1)}(t_0) \\ \rho^{(k+1)}(t_0) \end{bmatrix} + V\begin{bmatrix} x^{(k+1)}(t_f) \\ \rho^{(k+1)}(t_f) \end{bmatrix} = c \tag{2.12}$$

where

$$U = \begin{bmatrix} I & 0 \\ 0 & 0 \end{bmatrix}, \qquad V = \begin{bmatrix} 0 & 0 \\ -F & I \end{bmatrix}, \qquad c = \begin{bmatrix} x^0 \\ 0 \end{bmatrix} \tag{2.13}$$

The above linearization is basically achieved by calculating the coefficient matrices using the results from the previous iteration step. The time varying matrices are given by

$$\widetilde{A}_{ij}^{(k)} = A_{ij} - \frac{1}{2}\left[(M_j R^{-1} B^T + B R^{-1} M_j^T)\rho^{(k)}(t)\right]_i, \quad i,j = 1,...,n$$

$$\widetilde{Q}_{ij}^{(k)} = Q_{ij} - \frac{1}{2}\rho^{(k)}(t)^T (M_i R^{-1} M_j^T + M_j R^{-1} M_i^T)\rho^{(k)}(t),$$
$$i,j = 1,...,n \tag{2.14}$$

$$\widetilde{B}^{(k)} R^{-1} \widetilde{B}^{(k)T} = \left(B + \left\{x^{(k)} M\right\}\right) R^{-1} \left(B + \left\{x^{(k)} M\right\}\right)^T$$
$$- \frac{1}{2}\left(\left\{x^{(k)} M\right\} R^{-1} B^T + B R^{-1} \left\{x^{(k)} M\right\}^T\right)$$

The convergence of the above algorithm to the solution of (2.10) was proved in (Hofer and Tibken, 1988).

In this section, we exploit the iterative scheme (2.11), comprising a sequence of linear two-point boundary value problems, in order to derive the solution for the optimal open-loop control of singularly perturbed bilinear systems. The solution is obtained in the spirit of the general theory of singular perturbations, namely the problem is decomposed and studied in slow and fast time scales. The open-loop optimal control of singularly perturbed linear systems was studied in (Su et al., 1992a) and (Wilde and Kokotovic, 1973). The approach taken in (Wilde and Kokotovic, 1973) is efficient for an $O(\epsilon)$ of accuracy. In (Su et al., 1992a) a recursive approach is obtained such that an arbitrary order of accuracy, $O(\epsilon^k)$, $k = 1, 2, 3, ...$, can be obtained. The importance of the results reported in (Su et al., 1992a) is in the fact that the stiffness of the singularly perturbed two-point boundary value problem is converted into the problem of an ill-defined system of linear algebraic equations. The latter problem is much easier to handle. The study in (Su et al., 1992a) was limited to the time invariant systems. In this section, we show that following the ideas of (Su et al., 1992a) we will be able to handle in the same manner the time varying singularly perturbed two-point boundary value problem, such that an arbitrary order of accuracy can be obtained, and that the stiffness of the original problem is replaced by an ill-defined system of linear algebraic equations of order $2n$.

The singularly perturbed bilinear control system under consideration is represented by

$$\begin{bmatrix} \dot{y} \\ \epsilon\dot{z} \end{bmatrix} = \begin{bmatrix} A_1 & A_2 \\ A_3 & A_4 \end{bmatrix} \begin{bmatrix} y \\ z \end{bmatrix} + \begin{bmatrix} B_1 \\ B_2 \end{bmatrix} u + \left\{ \begin{bmatrix} y \\ z \end{bmatrix} \begin{bmatrix} M_s \\ M_f \end{bmatrix} \right\} u \qquad (2.15)$$

with initial conditions

$$\begin{bmatrix} y(t_0) \\ z(t_0) \end{bmatrix} = \begin{bmatrix} y^0 \\ z^0 \end{bmatrix}$$

where $y \in \Re^{n_1}$, $z \in \Re^{n_2}$ are, respectively, slow and fast state variables, ϵ is a small positive parameter, and

$$\left\{ \begin{bmatrix} y \\ z \end{bmatrix} \begin{bmatrix} M_s \\ M_f \end{bmatrix} \right\} = \sum_{j=1}^{n_1} y_j \begin{bmatrix} M_{sj} \\ M_{fj} \end{bmatrix} + \sum_{j=n_1+1}^{n_1+n_2} z_j \begin{bmatrix} M_{sj} \\ M_{fj} \end{bmatrix} \qquad (2.16)$$

A quadratic cost functional associated with (2.13) has the form

$$J = \frac{1}{2} \int_{t_0}^{t_f} \left(\begin{bmatrix} x \\ z \end{bmatrix}^T Q \begin{bmatrix} x \\ z \end{bmatrix} + u^T R u \right) dt \qquad (2.17)$$

The following notation is used in order to relate expressions in (2.4)-(2.6) and (2.13)-(2.15)

$$A = \begin{bmatrix} A_1 & A_2 \\ \frac{1}{\epsilon}A_3 & \frac{1}{\epsilon}A_4 \end{bmatrix}, \qquad Q = \begin{bmatrix} Q_1 & Q_2 \\ Q_2^T & Q_3 \end{bmatrix}$$

$$(2.18)$$

$$B = \begin{bmatrix} B_1 \\ \frac{1}{\epsilon}B_2 \end{bmatrix}, \quad M = \begin{bmatrix} M_s \\ \frac{1}{\epsilon}M_f \end{bmatrix}, \quad x(t) = \begin{bmatrix} y(t) \\ z(t) \end{bmatrix}$$

In the following, we will utilize the recursive scheme of (Hofer and Tibken, 1988), presented in (2.11)-(2.14), in order to find the optimal open-loop control law for the singularly perturbed bilinear-quadratic optimal control problem given by (2.15)-(2.18) in terms of the reduced-order slow and fast subsystems consistently in the spritit of the slow-fast time scale system decomposition of (Kokotovic et al., 1986).

It can be shown that the system of equations (2.11) preserves the singularly perturbed structure. Namely, the use of (2.15)-(2.18) in (2.7)-(2.8) and (2.11)-(2.14) produces

$$\begin{bmatrix} \dot{y} \\ \epsilon \dot{z} \end{bmatrix}^{(k+1)} = \widetilde{A}^{(k)} \begin{bmatrix} y \\ z \end{bmatrix}^{(k+1)} - \widetilde{B}^{(k)} R^{-1} \widetilde{B}^{(k)^T} \begin{bmatrix} \rho_s \\ \rho_f \end{bmatrix}^{(k+1)}$$

$$\begin{bmatrix} \dot{\rho}_s \\ \epsilon \dot{\rho}_f \end{bmatrix}^{(k+1)} = -\widetilde{Q}^{(k)} \begin{bmatrix} y \\ z \end{bmatrix}^{(k+1)} - \widetilde{A}^{(k)^T} \begin{bmatrix} \rho_s \\ \rho_f \end{bmatrix}^{(k+1)}$$

(2.19)

where $\rho_s \in \Re^{n_1}$ and $\rho_f \in \Re^{n_2}$ are costate vectors corresponding, respectively, to the slow variables $y^{(k+1)}$ and the fast variables $z^{(k+1)}$. The time varying matrices in (2.19) are given by

$$\widetilde{A}^{(k)} = \begin{bmatrix} \widetilde{A}_1^{(k)} & \widetilde{A}_2^{(k)} \\ \frac{1}{\epsilon} \widetilde{A}_3^{(k)} & \frac{1}{\epsilon} \widetilde{A}_4^{(k)} \end{bmatrix}, \qquad \widetilde{Q}^{(k)} = \begin{bmatrix} \widetilde{Q}_1^{(k)} & \widetilde{Q}_2^{(k)} \\ \widetilde{Q}_2^{(k)^T} & \widetilde{Q}_3^{(k)} \end{bmatrix}$$

$$\widetilde{S}^{(k)} = \widetilde{B}^{(k)} R^{-1} \widetilde{B}^{(k)^T} = \begin{bmatrix} \widetilde{S}_1^{(k)} & \frac{1}{\epsilon} \widetilde{S}_2^{(k)} \\ \frac{1}{\epsilon} \widetilde{S}_2^{(k)^T} & \frac{1}{\epsilon^2} \widetilde{S}_3^{(k)} \end{bmatrix}$$

(2.20)

After some algebra the state-costate equation (2.19) can be put in the standard singularly perturbed form

$$\begin{bmatrix} \dot{w}^{(k+1)} \\ \dot{\lambda}^{(k+1)} \end{bmatrix} = \begin{bmatrix} \widetilde{T}_1^{(k)} & \widetilde{T}_2^{(k)} \\ \frac{1}{\epsilon} \widetilde{T}_3^{(k)} & \frac{1}{\epsilon} \widetilde{T}_4^{(k)} \end{bmatrix} \begin{bmatrix} w^{(k+1)} \\ \lambda^{(k+1)} \end{bmatrix}$$

(2.21)

where the new notation is

$$w^{(k+1)} = \begin{bmatrix} y^{(k+1)} \\ \rho_s^{(k+1)} \end{bmatrix}, \quad \lambda^{(k+1)} = \begin{bmatrix} z^{(k+1)} \\ \rho_f^{(k+1)} \end{bmatrix}, \quad \rho^{(k+1)} = \begin{bmatrix} \rho_s^{(k+1)} \\ \rho_f^{(k+1)} \end{bmatrix} \quad (2.22)$$

The time varying matrices $\widetilde{T}_i^{(k)}$ introduced in (2.21) have the following forms

$$\widetilde{T}_1^{(k)} = \begin{bmatrix} \widetilde{A}_1^{(k)} & -\widetilde{S}_1^{(k)} \\ -\widetilde{Q}_1^{(k)} & -\widetilde{A}_1^{(k)^T} \end{bmatrix}, \quad \widetilde{T}_2^{(k)} = \begin{bmatrix} \widetilde{A}_2^{(k)} & -\widetilde{S}_2^{(k)} \\ -\widetilde{Q}_2^{(k)} & -\widetilde{A}_3^{(k)^T} \end{bmatrix}$$

$$\widetilde{T}_3^{(k)} = \begin{bmatrix} \widetilde{A}_3^{(k)} & -\widetilde{S}_2^{(k)^T} \\ -\widetilde{Q}_2^{(k)^T} & -\widetilde{A}_2^{(k)^T} \end{bmatrix}, \quad \widetilde{T}_4^{(k)} = \begin{bmatrix} \widetilde{A}_4^{(k)} & -\widetilde{S}_3^{(k)} \\ -\widetilde{Q}_3^{(k)} & -\widetilde{A}_4^{(k)^T} \end{bmatrix}$$

(2.23)

The expression for the boundary conditions is changed due to an interchange of rows corresponding to $\rho_s^{(k+1)}$ and $z^{(k+1)}$, which modifies matrices defined in (2.12) as follows

$$U_1 \begin{bmatrix} w^{(k+1)}(t_0) \\ \lambda^{(k+1)}(t_0) \end{bmatrix} + V_1 \begin{bmatrix} w^{(k+1)}(t_f) \\ \lambda^{(k+1)}(t_f) \end{bmatrix} = c_1 \qquad (2.24)$$

where

$$U_1 = \begin{bmatrix} I_{n_1} & 0 & 0 & 0 \\ 0 & 0 & 0 & 0 \\ 0 & 0 & I_{n_2} & 0 \\ 0 & 0 & 0 & 0 \end{bmatrix}, \qquad c_1 = \begin{bmatrix} y^0 \\ 0 \\ z^0 \\ 0 \end{bmatrix}$$

and

$$V_1 = \begin{bmatrix} 0 & 0 & 0 & 0 \\ -F_1 & I_{n_1} & -\epsilon F_2 & 0 \\ 0 & 0 & 0 & 0 \\ -F_2^T & 0 & -F_3 & I_{n_2} \end{bmatrix} \qquad (2.25)$$

In order to obtain the slow and fast decoupled subsystems from (2.21), we apply the Chang transformation (Chang, 1972), see Appendix 2.1. In this section, we use a new version of the Chang transformation obtained by (Qureshi and Gajic, 1992), which is given by

$$\mathbf{T}_1^{(\mathbf{k})}(t, \epsilon) = \begin{bmatrix} I_1 & -\epsilon H^{(k)} \\ -L^{(k)} & I_2 \end{bmatrix}$$

$$\mathbf{T}_1^{(\mathbf{k})^{-1}}(t, \epsilon) = \begin{bmatrix} I_1 - \epsilon H^{(k)} W^{(k)} L^{(k)} & \epsilon H^{(k)} W^{(k)} \\ W^{(k)} L^{(k)} & W^{(k)} \end{bmatrix} \qquad (2.26)$$

with

$$W^{(k)} = \left(I_2 - \epsilon L^{(k)} H^{(k)} \right)^{-1} \qquad (2.27)$$

where I_1 and I_2 are identity matrices of order $2n_1$ and $2n_2$, respectively. The matrices $H^{(k)}$ and $L^{(k)}$ are the solutions of the following completely decoupled stiff matrix differential equations

$$\epsilon \dot{H}^{(k)} = -H^{(k)} \tilde{T}_4^{(k)} + \tilde{T}_2^{(k)} + \epsilon \left(\tilde{T}_1^{(k)} H^{(k)} - H^{(k)} \tilde{T}_3^{(k)} H^{(k)} \right)$$

$$\epsilon \dot{L}^{(k)} = \tilde{T}_4^{(k)} L^{(k)} + \tilde{T}_3^{(k)} + \epsilon \left(L^{(k)} \tilde{T}_1^{(k)} + L^{(k)} \tilde{T}_2^{(k)} L^{(k)} \right) \qquad (2.28)$$

The initial conditions for differential equations (2.28) are arbitrary (Chang, 1972) so that the boundary layers can be eliminated by properly choosing the initial conditions as

$$H^{(k)}(t_0) = \tilde{T}_2^{(k)}(t_0)\tilde{T}_4^{(k)-1}(t_0)$$

$$L^{(k)}(t_0) = -\tilde{T}_4^{(k)-1}(t_0)\tilde{T}_3^{(k)}(t_0)$$

The existence of the solutions of (2.28) for sufficiently small values of ϵ is established in (Qureshi, 1992; Qureshi and Gajic, 1992).

The transformation (2.26) applied to the system (2.21) produces two completely decoupled subsystems

$$\dot{\eta}^{(k)} = \left(\tilde{T}_1^{(k)} - H^{(k)}\tilde{T}_3^{(k)}\right)\eta^{(k)} \tag{2.29}$$

$$\dot{\xi}^{(k)} = \frac{1}{\epsilon}\left(\tilde{T}_4^{(k)} - \epsilon L^{(k)}\tilde{T}_2^{(k)}\right)\xi^{(k)} \tag{2.30}$$

with

$$\begin{bmatrix} \eta^{(k)} \\ \xi^{(k)} \end{bmatrix} = \mathbf{T}_1^{(k)}(t,\epsilon)\begin{bmatrix} w^{(k)} \\ \lambda^{(k)} \end{bmatrix} \tag{2.31}$$

Consequently, the change of variables transforms the boundary conditions

$$U_2\begin{bmatrix} \eta^{(k)}(t_0) \\ \xi^{(k)}(t_0) \end{bmatrix} + V_2\begin{bmatrix} \eta^{(k)}(t_f) \\ \xi^{(k)}(t_f) \end{bmatrix} = c_1 \tag{2.32}$$

where

$$U_2 = U_1\mathbf{T}_1^{(k)-1}(t_0,\epsilon), \qquad V_2 = V_1\mathbf{T}_1^{(k)-1}(t_f,\epsilon) \tag{2.33}$$

The solutions of equations (2.29) and (2.30) are

$$\eta^{(k)}(t) = \Phi^{(k)}(t,t_0,\epsilon)\eta^{(k)}(t_0)$$

$$\xi^{(k)}(t) = \Psi^{(k)}(t,t_0,\epsilon)\xi^{(k)}(t_0) \tag{2.34}$$

where $\Phi(t,t_0,\epsilon)$ and $\Psi(t,t_0,\epsilon)$ are the transition matrices of (2.29) and (2.30), respectively.

The initial conditions $\eta^{(k)}(t_0)$ and $\xi^{(k)}(t_0)$ have to be determined. This can be done as follows. Substitution of (2.34) into (2.32) yields

$$\Delta(\epsilon) \begin{bmatrix} \eta^{(k)}(t_0) \\ \xi^{(k)}(t_0) \end{bmatrix} = c_1 \tag{2.35}$$

where

$$\Delta(\epsilon) = U_2(\epsilon) + V_2(\epsilon) \begin{bmatrix} \Phi(t_f, t_0, \epsilon) & 0 \\ 0 & \Psi(t_f, t_0, \epsilon) \end{bmatrix} \tag{2.36}$$

If $\Delta^{-1}(\epsilon)$ exists then the solution of (2.35) will be

$$\begin{bmatrix} \eta^{(k)}(t_0) \\ \xi^{(k)}(t_0) \end{bmatrix} = \Delta^{-1}(\epsilon) c_1 \tag{2.37}$$

Note that as $\epsilon \to 0$

$$\left\{ T_1^{(k)}(t, 0) \right\}^{-1} = \begin{bmatrix} I_1 & 0 \\ L^{(k)}(t_0) & I_2 \end{bmatrix} \tag{2.38}$$

and therefore

$$U_2 = U_1 \begin{bmatrix} I_1 & 0 \\ L^{(k)}(t_0) & I_2 \end{bmatrix}, \quad V_2 = V_1 \begin{bmatrix} I_1 & 0 \\ L^{(k)}(t_f) & I_2 \end{bmatrix} \tag{2.39}$$

After partitioning the transition matrices $\Phi(t, t_0, 0)$ and $\Psi(t, t_0, 0)$ as

$$\Phi(t, t_0, 0) = \begin{bmatrix} \Phi_{11}(t, t_0, 0) & \Phi_{12}(t, t_0, 0) \\ \Phi_{21}(t, t_0, 0) & \Phi_{22}(t, t_0, 0) \end{bmatrix}$$

$$\Psi(t, t_0, 0) = \begin{bmatrix} \Psi_{11}(t, t_0, 0) & \Psi_{12}(t, t_0, 0) \\ \Psi_{21}(t, t_0, 0) & \Psi_{22}(t, t_0, 0) \end{bmatrix}$$

and after some algebra the matrix $\Delta(\epsilon)$ is obtained in the form

$$\Delta(\epsilon) = \begin{bmatrix} I_{n_1} & 0 & 0 & 0 \\ * & \Delta_{22} & 0 & 0 \\ * & * & I_{n_2} & 0 \\ * & * & * & \Delta_{44} \end{bmatrix} + O(\epsilon) \tag{2.40}$$

where

$$\Delta_{22} = \Phi_{22}(t_f, t_0, 0) - F_1 \Phi_{12}(t_f, t_0, 0)$$

$$\Delta_{44} = \Psi_{22}(t_f, t_0, 0) - F_3 \Psi_{12}(t_f, t_0, 0)$$

The asterisks denote the terms that are not important for the nonsingularity of $\Delta(\epsilon)$.

Since the matrices Δ_{22} and Δ_{44} are nonsingular (Kirk, 1970, page 211), so does $\Delta(\epsilon)$ for sufficiently small values of ϵ, with $0 < \epsilon \leq \epsilon_1$ and ϵ sufficiently small.

Note that due to presence of the $\frac{1}{\epsilon}$ term in $\Psi(t_f, t_0, 0)$, the system of linear algebraic equations (2.35) is ill-conditioned. However, this problem is much easier to be solved than the original two-point stiff boundary value problem. In summary, we have established the following theorem.

Theorem 2.1 *Let the problem matrices be continuous functions of t on the time interval $t_0 \leq t \leq t_f$, then for all sufficiently small ϵ the boundary value problem (2.29)–(2.30) and (2.33) has the solution given by*

$$\begin{bmatrix} \eta(t, \epsilon) \\ \xi(t, \epsilon) \end{bmatrix} = \begin{bmatrix} \Phi(t, t_0, \epsilon) & 0 \\ 0 & \Psi(t, t_0, \epsilon) \end{bmatrix} \Delta^{-1}(\epsilon) c_1$$

∎

Consequently, the solution of the boundary problem (2.21)-(2.24) is obtained as

$$\begin{bmatrix} w^{(k+1)}(t, \epsilon) \\ \lambda^{(k+1)}(t, \epsilon) \end{bmatrix} = \left\{ \mathbf{T}_1^{(k)}(t, \epsilon) \right\}^{-1} \begin{bmatrix} \eta^{(k+1)}(t, \epsilon) \\ \xi^{(k+1)}(t, \epsilon) \end{bmatrix} \tag{2.41}$$

so that the required variables $y^{(k+1)}$ and $z^{(k+1)}$ can be found by partitioning the vectors $w^{(k+1)}$ and $\lambda^{(k+1)}$ according to (2.22).

The main problem that we are faced with in the presented method is the problem of finding the transition matrices $\Phi(t, t_0, \epsilon)$ and $\Psi(t, t_0, \epsilon)$ of the corresponding time varying systems. One way to overcome this problem is to study the optimal open-loop control of singularly perturbed system in the discrete-time domain, where the analytic expressions for the corresponding transition matrices exist. In that direction the first step should be to develop the discrete-time version of the results of (Hofer and Tibken, 1988). Note that the solution to the time invariant version of the open-loop optimal control of discrete-time singularly perturbed systems is obtained in (Qureshi et al., 1991) so that the second step would be to develop the time varying version of the results of (Qureshi et al., 1991). The above indicates an important research problems for future studies. In addition of clarifying some theoretical aspects, the problem representation in the discrete-time domain might simplify computational requirements as well.

2.4 "Closed-Loop" Optimal Control of Singularly Perturbed Bilinear Systems

Consider the optimal control problem of a bilinear system represented by (2.4)-(2.6). The closed-loop solution of the optimization problem (2.4)-(2.5) at steady state ($t_f = \infty$) yields to the optimal control in the form

$$u_{opt} = -R^{-1}(B + \{xM\})^T P(x)x \qquad (2.42)$$

where $P(x)$ is the solution of the following equation (Cebuhar and Constanza, 1984)

$$Q + P(x)A + A^T P(x)$$
$$-P(x)(B + \{xM\})R^{-1}(B + \{xM\})^T P(x) = 0 \qquad (2.43)$$

This nonlinear system of algebraic matrix equations is very hard to solve. In general it has no analytical solution. However, it has been shown in (Cebuhar and Constanza, 1984) that the approximate solution can be obtained from the sequence of linear systems

$$\dot{x}_0 = Ax_0 + Bu, \qquad x_0(t_0) = x^0$$
$$\dot{x}_i = Ax_i + B_i(t)u_i, \qquad B_i(t) \triangleq B + \{x_{i-1}(t)M\}, \quad x_i(t_0) = x^0 \qquad (2.44)$$
$$i = 1, 2, 3, \ldots$$

and the sequence of the time varying algebraic Riccati equations

$$Q + P_i(t)A + A^T P_i(t) - P_i(t)B_i(t)R^{-1}B_i^T(t)P_i(t) = 0 \qquad (2.45)$$

Solutions of (2.44) and (2.45) produce the sequence of the approximate "feedback" controls

$$u_i^*(t) = -R^{-1}B_i^T(t)P_i(t)x_i(t) = -R^{-1}[B + \{x_{i-1}(t)M\}]P_i(t)x_i(t)$$
$$(2.46)$$

such that

$$u_i^*(t) \to u_{opt}(t), \quad x_i^*(t) \to x_{opt}(t) \tag{2.47}$$

with

$$u_i^*(t) \to u_\infty^*(t) = -R^{-1}BP_\infty x_\infty(t) - R^{-1}B\{x_\infty(t)M\}P_\infty x_\infty(t)$$

This expression indicates the quadratic nature of the optimal "closed-loop" control for bilinear systems. The convergence stated in (2.47) is uniform in t, and is guaranteed under the following assumption.

Assumption 2.1 The pair (A, B) is controllable and x stays in the controllability domain $X_c = \{x \in R^n | (A, B + \{xM\})$ *controllable*$\}$.

$\triangle$

Note that the algebraic Riccati equation whose coefficient matrices are functions of a parameter t is studied in (Ran and Rodman, 1988) for the case $t \in (t_0, t_f)$.

It is important to point out that (2.45) and (2.46) establish in some sense the optimal linear feedback law. Namely, using the feedback coefficient from (2.45) in the linear feedback loop around the bilinear system (2.4) and feeding back the state variables multiplied by these coefficients produces the approximate linear feedback law. This is a very strong result since it is known that it is impossible to get, in general, the optimal feedback control of nonlinear (and thus bilinear) systems due to fact that the partial differential Hamilton-Bellman-Jacobi equation has no analytical solutions.

In this chapter, we will relax the controllability assumption into the stabilizability assumption (Kucera, 1972). Also, since the matrix Q in (2.45) does not change per iteration it is convenient to assume that the pair $(A, \sqrt{Q})$ is detectable. This will establish the existence of the unique stabilizing solution $P_i(t)$, in order words, the matrix $A - B_i(t)R^{-1}B_i(t)P_i(t)$ will be stable for every frozen $t \in [0, \infty)$. Due to stability of the closed-loop system matrix, at steady state we have $0 = \left(A - B_i(t)R^{-1}B_i(t)P_i(t)\right)x_e(t)$, that is, the unique equilibrium point of the "bilinear system" is the origin, so that $B_i(t) \rightarrow B = const$, and the equation (2.45) tends to the time-invariant algebraic Riccati equation. The required optimal feedback control (2.46) in that case tends to zero, so there is no need to solve the equation (2.45) over an infinite period of time. Thus, we will use the following assumption.

Assumption 2.2 The pair (A, B) is stabilizable, x stays in the stabilizability domain $X_s = \{x \in R^n | (A, B + \{xM\})$ $stabilizable\}$, and the pair $(A, \sqrt{Q})$ is detectable.

$$\triangle$$

The main goal of this section is to exploit the iterative procedure (2.44)-(2.46) for the singularly perturbed bilinear structure (2.15)-(2.18) in order to get an expression for the near-optimal control in terms of the reduced-order slow and fast subsystems (Kokotovic et al., 1986). There are two important reasons for this study: 1) to avoid an ill-defined numerical problem associated with the equation (2.45) subject to (2.18); and, 2) to reduce the size of required computations and generate the near-optimal solution in parallel — in slow and fast time scales, and speed up the optimization process.

2.4.1 Composite Control of Bilinear Singularly Perturbed Systems

The composite control of singularly perturbed linear systems is derived in (Chow and Kokotovic, 1976). In that paper it is shown how to decompose the linear-quadratic optimal control problem of singularly perturbed systems into the corresponding slow and fast reduced-order linear-quadratic optimal control problems with the accuracy of $O(\epsilon)$. Optimizing independently the slow and fast subsystems the composite control, close an $O(\epsilon)$ to the optimal one, is obtained as a linear combination of the slow and fast subsystem optimal controls, that is

$$u_c = u_s^{opt} + u_f^{opt} = u^{opt} + O(\epsilon)$$

Following the result of (Chow and Kokotovic, 1976), the composite control of the sequence of the linear-quadratic optimal control problems (2.44)-(2.47), subject to the singularly perturbed structure (2.15)-(2.18), can be obtained from the slow and fast time scales linear-quadratic optimal control problems. Note that on the contrary to (Chow and Kokotovic, 1976), we are faced with the time varying problem. The slow time scale problem of order n_1 for the block diagonal structure of the penalty matrix Q (it has been assumed without loss of generality that $Q_2 = 0$), is given by

$$\dot{y}_s = A_0 y_s + B_s(t) u_s, \quad y_s(t_0) = y^0$$

$$z_s(t) = -A_4^{-1}(A_3 y_s + B_{2i}(t) u_s) \tag{2.48}$$

$$J_s = \frac{1}{2} \int_{t_0}^{\infty} (y_s^T Q_0 y_s + 2 u_s^T D_s y_s + u_s^T R_s u_s) \, dt$$

where

$$B_i(t) = B + \{x_{i-1}(t)M\} = \begin{bmatrix} B_{1i}(t) \\ \frac{1}{\epsilon} B_{2i}(t) \end{bmatrix}$$

$$A_0 \triangleq A_1 - A_2 A_4^{-1} A_3, \quad B_s(t) \triangleq B_{1i}(t) - A_2 A_4^{-1} B_{2i}(t)$$

$$Q_0 = Q_1 + A_3^T A_4^{-T} Q_3 A_4^{-1} A_3 \tag{2.49}$$
$$D_s(t) = B_{2i}^T(t) A_4^{-T} Q_3 A_4^{-1} A_3$$

$$R_s(t) = R + B_{2i}^T(t) A_4^{-T} Q_3 A_4^{-1} B_{2i}(t)$$

The optimal slow control strategy is

$$u_s(t) = -R_s^{-1}(t)\left(D_s(t) + B_s^T(t) P_s(t)\right) y_s(t) = G_0(t) y_s(t) \tag{2.50}$$

where $P_s(t)$ satisfies the algebraic Riccati equation

$$P_s(t)A_s(t) + A_s^T(t)P_s(t) + Q_s(t) - P_s(t)B_s(t)R_s^{-1}(t)B_s^T(t)P_s(t) = 0 \tag{2.51}$$

with

$$A_s(t) = \left(A_0(t) - B_s(t)R_s^{-1}(t)D_s(t)\right)$$
$$Q_s(t) = Q_0 - D_s^T(t)R_s^{-1}(t)D_s(t) \tag{2.52}$$

The fast time scale optimization problem of order n_2 is given by

$$\epsilon\dot{z}_f = A_4 z_f + B_{2i}(t)u_f, \quad z_f(t_0) = z^0 - z_s(t_0)$$

$$\tag{2.53}$$

$$J_f = \frac{1}{2}\int_{t_0}^{\infty} (z_f^T Q_2 z_f + u_f^T R u_f)\,dt$$

where $z_f = z - z_s$ and $u_f = u - u_s$ denote fast parts of the corresponding variables. The optimal control for the fast subsystem is

$$u_f(t) = -R^{-1}B_{2i}^T(t)P_f(t)z_f(t) = G_2(t)z_f(t) \tag{2.54}$$

where $P_f(t)$ is the solution of the "fast" algebraic Riccati equation

$$P_f(t)A_4 + A_4^T P_f(t) - P_f(t)B_{2i}(t)R^{-1}B_{2i}^T(t)P_f(t) + Q_3 = 0 \tag{2.55}$$

A realizable composite control requires that the system states y_s and z_f be expressed in terms of the actual system states y and z. This can be achieved by replacing y_s by y and z_f by $z - z_s$ so that

$$u_c(t) = G_2(t)\left[z(t) + A_4^{-1}(A_3 y(t) + B_{2i}(t)G_0(t)y(t))\right] + G_0(t)y$$
$$= G_1(t)y(t) + G_2(t)z(t) \tag{2.56}$$

where

$$G_1 = \left(I_r + G_2 A_4^{-1}B_{2i}\right)G_0 + G_2 A_4^{-1}A_3 \tag{2.57}$$

Thus, instead of solving at each iteration the global full-order numerically ill-defined algebraic Riccati equation (2.45), in the presented slow-fast decomposition technique we are faced with the problem of solving two reduced-order well-defined algebraic Riccati equations (2.51) and (2.55).

The unique solutions of (2.51) and (2.55) exist under the following assumption, which follows from the works of (Kokotovic and Yackel, 1972; Cebuhar and Constanza, 1984).

Assumption 2.3 The pairs $(A_s(t), B_s(t))$ and (A_4, B_{2i}) are stabilizable, the pairs $\left(A_s(t), \sqrt{Q_s(t)}\right)$ and $(A_4, \sqrt{Q_3})$ are detectable, and $x_i(t)$ stay in the stabilizability domains of the slow and fast subsystems for every $t \geq t_0$.

$$\triangle$$

The near optimality of the composite control law is stated in the following lemma.

Lemma 2.1 *Under Assumption 2.3 the composite control law (2.56) is suboptimal in the sense*

$$u_{opt}(t) = u_c(t) + O(\epsilon), \qquad t \geq t_0$$

$$y(t) = y_s(t) + O(\epsilon), \qquad t \geq t_0$$
$$z(t) = z_s(t) + z_f(t) + O(\epsilon), \qquad t \geq t_0$$

(2.58)

The proof of this lemma follows from (Chow and Kokotovic, 1976; Kokotovic et al., 1986).

An $O(\epsilon)$ perturbation in each iteration of the presented slow-fast iterative scheme given by (2.57) will propagate into the next iteration, but due to the continuous dependence of the solution of the sequence of linear differential equations (2.44) with respect to perturbations in system coefficients, the presented method produces

$$u_{ci}^*(t) \rightarrow u^*(t) + O(\epsilon)$$

$$\begin{bmatrix} y_{si}^*(t) \\ z_{si}^*(t) + z_{fi}^*(t) \end{bmatrix} \rightarrow x^*(t) + O(\epsilon)$$

(2.59)

where i stands for the iteration number.

The efficiency of the proposed composite control suboptimal procedure is demonstrated in the next section on a real physical bilinear-quadratic control problem.

2.5 Case Study: Induction Motor Drives

Consider a fourth-order example representing the real physical model of
induction motor drives (Figalli et al., 1984). A frequency controlled two
phase induction motor can be put in the bilinear singularly perturbed form
(2.15). The state and control variable are

$$x = \begin{bmatrix} y_1 \\ y_2 \\ z_1 \\ z_2 \end{bmatrix} = \begin{bmatrix} \phi_{ds} \\ \phi_{qs} \\ i_{ds} \\ i_{qs} \end{bmatrix}, \quad u = \begin{bmatrix} u_1 \\ u_2 \\ u_3 \end{bmatrix} = \begin{bmatrix} v_{ds} \\ v_{qs} \\ \omega_s \end{bmatrix}$$

where

ϕ_{ds} and ϕ_{qs} — projections of the stator flux
i_{ds} and i_{qs} — projections of stator current
v_{ds} and v_{qs} — projections of the supply voltage
ω_s — slip angular frequency.

The problem matrices have the following values

$$A = \begin{bmatrix} 0 & 321.57 & -0.312 & 0 \\ -312.57 & 0 & 0 & -0.312 \\ 98.87 & 27059 & -44.93 & 2.57 \\ -27059 & 98.87 & -2.57 & -44.93 \end{bmatrix}$$

$$B = \begin{bmatrix} 1 & 0 & 0 \\ 0 & 1 & -7.3 \\ 87.3 & 0 & 87.8 \\ 0 & 87.3 & -53 \end{bmatrix}, \quad x(t_0) = \begin{bmatrix} -0.07 \\ 0.04 \\ 15 \\ 47 \end{bmatrix}$$

$$N_1 = \begin{bmatrix} 0 & 0 & 0 \\ 0 & 0 & -1 \\ 0 & 0 & 0 \\ 0 & 0 & 0 \end{bmatrix}, \quad N_2 = \begin{bmatrix} 0 & 0 & 1 \\ 0 & 0 & 0 \\ 0 & 0 & 0 \\ 0 & 0 & 0 \end{bmatrix}, \quad N_3 = \begin{bmatrix} 0 & 0 & 0 \\ 0 & 0 & 0 \\ 0 & 0 & 0 \\ 0 & 1 & 0 \end{bmatrix}$$

$$N_4 = \begin{bmatrix} 0 & 0 & 0 \\ 0 & 0 & 0 \\ 0 & 0 & 1 \\ 0 & 0 & 0 \end{bmatrix}, \quad Q = I_4, \quad R = \begin{bmatrix} 0.1 & 0 & 0 \\ 0 & 0.1 & 0 \\ 0 & 0 & 50 \end{bmatrix}$$

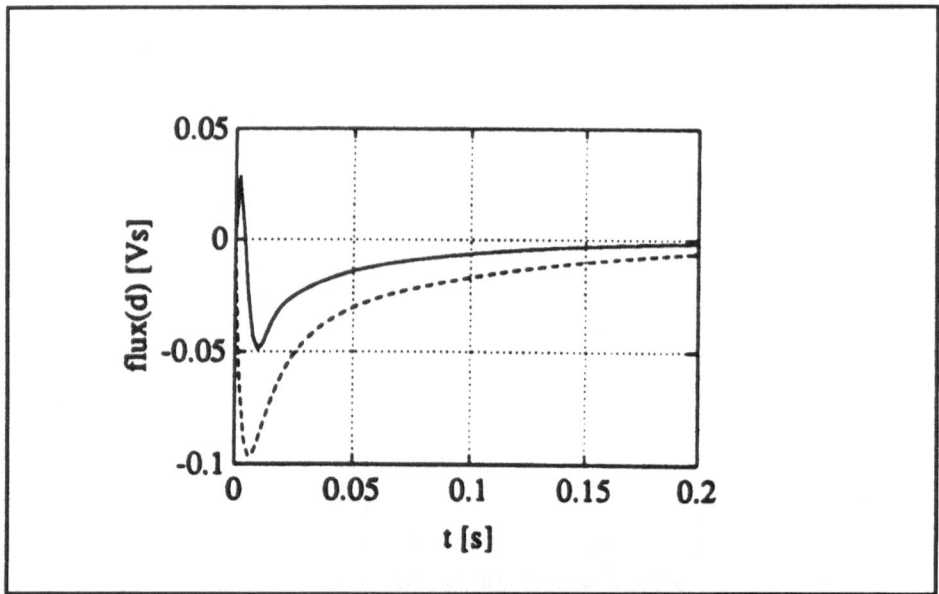

Figure 2.1: Optimal and approximate trajectories for flux ϕ_{ds}

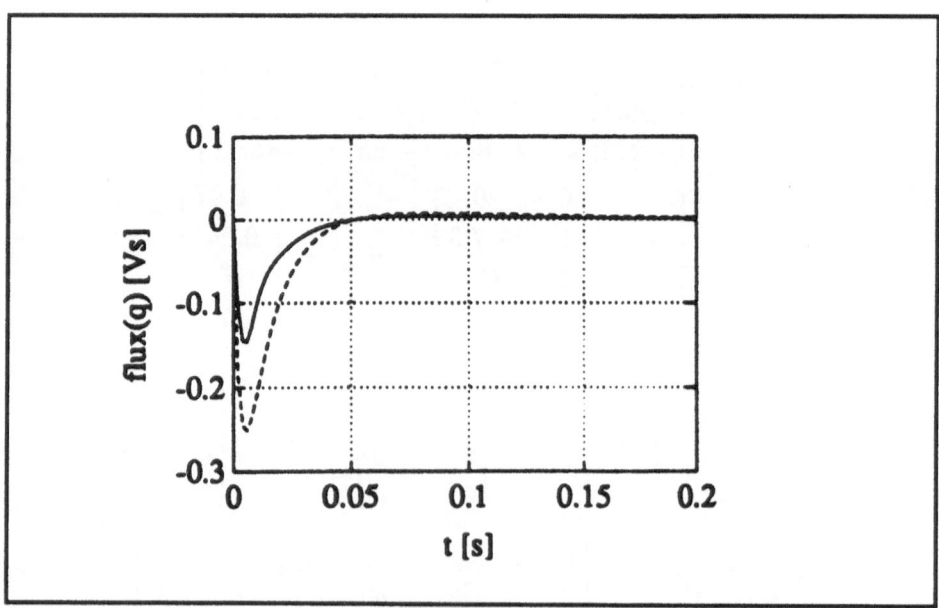

Figure 2.2: Optimal and approximate trajectories for flux ϕ_{qs}

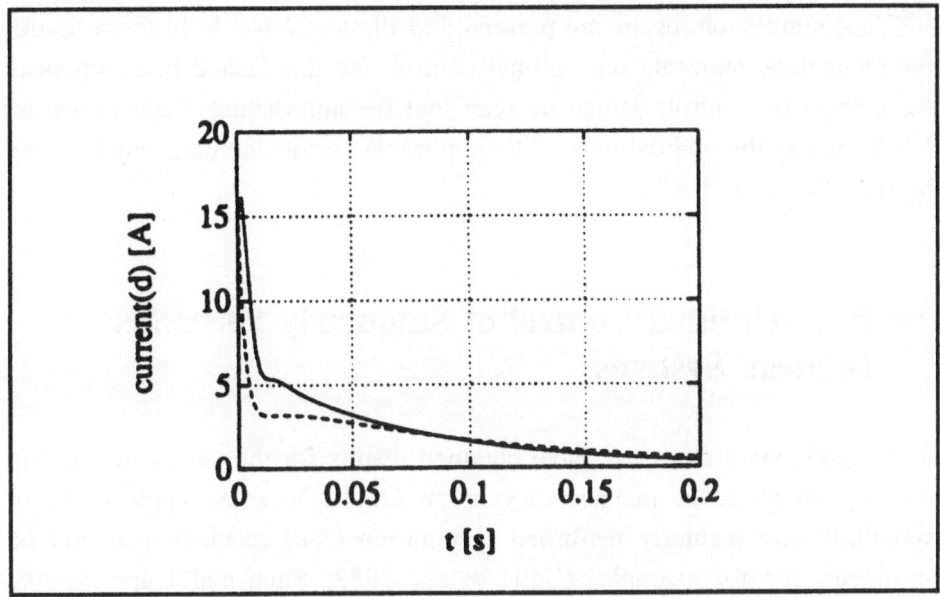

Figure 2.3: Optimal and approximate trajectories for current i_{ds}

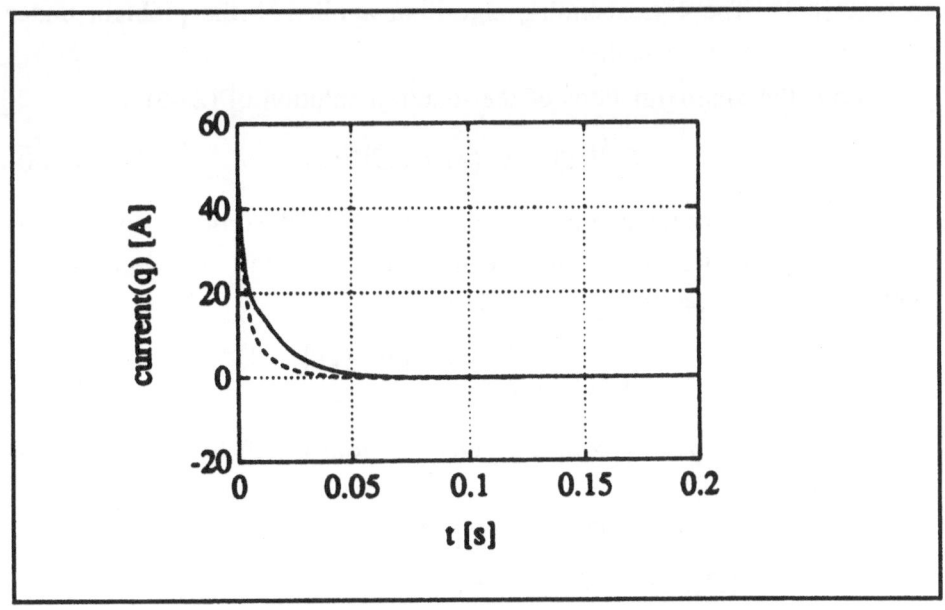

Figure 2.4: Optimal and approximate trajectories for current i_{qs}

The simulation results are presented in Figures 2.1–2.4. In these figures the solid lines represent the optimal control and the dashed lines represent the composite control. It can be seen that the approximate trajectories are $O(\epsilon)$ close to the optimal ones. All simulations results are obtained by using MATLAB, (Hill, 1988).

2.6 Near-Optimal Control of Singularly Perturbed Bilinear Systems

In the previous section, we have obtained results for the composite control law, which produces the accuracy of an $O(\epsilon)$. In some applications of singularly and regularly perturbed systems an $O(\epsilon)$ accuracy may not be sufficient, see for example, (Gajic et al., 1989; Shen and Gajic, 1990a; Skataric and Gajic, 1992). The iterative refinement of (Gajic et al., 1990), based on the fixed point iterations, to be performed at each discrete-time instant along time axis, can be used to increase the accuracy in (2.58)-(2.59) up to $O(\epsilon^k)$. The corresponding algorithm applied to the problem under consideration is given bellow.

Define the approximations of the required solution of (2.45) as

$$P_i^{(k)}(\epsilon) = \mathbf{P}_i(\epsilon) + \epsilon E_i^{(k)}(\epsilon) \tag{2.60}$$

where $E_i^{(k)}$ stand for the approximation errors, k for the order of approximation, and i is the iteration number with respect to (2.45). The zeroth-order solution, $\mathbf{P}_i(t)$, is partitioned according to

$$\mathbf{P_i}(t) = \begin{bmatrix} \mathbf{P_{1i}}(t) & \epsilon\mathbf{P_{2i}}(t) \\ \epsilon\mathbf{P_{2i}^T}(t) & \epsilon\mathbf{P_{3i}}(t) \end{bmatrix} \tag{2.61}$$

The elements $\mathbf{P}_{ji}$, $j = 1, 2, 3$, are obtained from (2.45) by setting $\epsilon = 0$, that is

$$\mathbf{P_{1i}}(t) = P_{si}(t)$$
$$\mathbf{P_{2i}}(t) = -\left(\mathbf{P_{1i}}A_2 + A_3^T\mathbf{P_{3i}} - \mathbf{P_{1i}}S_i\mathbf{P_{3i}}\right)\left(A_4 - S_{2i}\mathbf{P_{3i}}\right)^{-1} \tag{2.62}$$
$$\mathbf{P_{3i}}(t) = P_{fi}(t)$$

Note that $P_{si}(t)$ and $P_{fi}(t)$ are obtained from (2.51) and (2.55), where i stands for the given iteration of (2.45), and newly defined matrices are

$$S_i = B_{1i}R^{-1}B_{2i}^T, \quad S_{ji} = B_{ji}R^{-1}B_{ji}^T, \quad j = 1,2 \qquad (2.63)$$

The approximation errors partitioned as

$$E_i^{(k)} = \begin{bmatrix} E_{1i}^{(k)} & \epsilon E_{2i}^{(k)} \\ \epsilon E_{2i}^{(k)^T} & \epsilon E_{3i}^{(k)} \end{bmatrix} \qquad (2.64)$$

can be obtained from the following fixed point algorithm (Gajic et al., 1990)

$$E_{3i}^{(k+1)}D_{3i} + D_{3i}^T E_{3i}^{(k+1)} = H_{3i}^{(k)}$$

$$E_{2i}^{(k+1)}D_{3i} + E_{1i}^{(k+1)}D_{21i} + D_{22i}^T E_{3i}^{(k+1)} = -H_{1i}^{(k)}$$

$$E_{1i}^{(k+1)}D_{1i} + D_{1i}^T E_{1i}^{(k+1)} = D^T H_{1i}^{(k)^T} + H_{1i}^{(k)}D + D_i^T H_{3i}^{(k)}D_i + \epsilon H_{2i}^{(k)}$$

$$E_{ji}^{(0)} = 0, \quad j = 1,2,3$$

$$(2.65)$$

where

$$D_{1i} = D_{11i} - D_{21i}D_{3i}^{-1}D_{22i}, \quad D_{11i} = A_1 - S_{1i}\mathbf{P}_{1i} - S_i\mathbf{P}_{2i}^{\mathbf{T}}$$
$$D_{21i} = A_2 - S_i\mathbf{P}_{3i}, \qquad\qquad D_{22i} = A_3 - S_i^T\mathbf{P}_{1i} - S_{2i}\mathbf{P}_{2i}^{\mathbf{T}} \quad (2.66)$$
$$D_{3i} = A_4 - S_{2i}\mathbf{P}_{3i}, \qquad\qquad D_i = D_{3i}^{-1}D_{22i}$$

and

$$H_{1i}^{(k)} = A_1^T P_{2i}^{(k)} - P_{1i}^{(k)}S_{1i}P_{2i}^{(k)} - P_{2i}^{(k)}S_i^T P_{2i}^{(k)}$$
$$-\epsilon\left(E_{1i}^{(k)}S_i E_{3i}^{(k)} + E_{2i}^{(k)}S_{2i}E_{3i}^{(k)}\right) \qquad (2.67)$$

$$H_{2i}^{(k)} = E_{1i}^{(k)}S_{1i}E_{1i}^{(k)} + E_{1i}^{(k)}S_i E_{2i}^{(k)^T} + E_{2i}^{(k)}S_i^T E_{1i}^{(k)} + E_{2i}^{(k)}S_{2i}E_{2i}^{(k)^T}$$

$$H_{3i}^{(k)} = -P_{2i}^{(k)^T} - A_2^T P_{2i}^{(k)} + \epsilon P_{2i}^{(k)^T}S_{1i}P_{2i}^{(k)} + \epsilon E_{3i}^{(k)}S_{2i}E_{3i}^{(k)}$$
$$+P_{2i}^{(k)^T}S_i P_{3i}^{(k)} + P_{3i}^{(k)}S_i^T P_{2i}^{(k)}$$

$$(2.68)$$

Note that matrices D_{1i} and D_{3i} are stable (Gajic et al., 1990). This algorithm converges with the rate of convergence of $O(\epsilon)$, that is (Gajic et al., 1990)

$$\left\| E_{ji}^{(k+1)} - E_{ji}^{(k)} \right\| = O(\epsilon), \quad j = 1, 2, 3 \tag{2.69}$$

or

$$\left\| E_{ji}^{(k+1)} - E_{ji} \right\| = O\left(\epsilon^k\right), \quad j = 1, 2, 3 \tag{2.70}$$

The approximation

$$P_{ji}^{(k)}(\epsilon) = \mathbf{P_{ji}}(\epsilon) + \epsilon E_{ji}^{(k)}(\epsilon), \quad j = 1, 2, 3 \tag{2.71}$$

will produce $O\left(\epsilon^{k+1}\right)$ approximation of the required solution P_i. Thus, having obtained P_i with the accuracy of $O\left(\epsilon^{k+1}\right)$, we get the same accuracy for the optimal trajectories and the approximate optimal control law. The price for this is that we have to solve k-times two reduced-order Lyapunov equations (2.65) at each discrete-time instant in the interval of interest. There are several methods available in the literature for solving the algebraic Lyapunov equation. A summary of these methods can be found in (Gajic and Qureshi, 1995).

In the recent paper (Rutkowski and Gajic, 1993; see also Rutkowski, 1995) a hybrid Newton/fixed-point iterations algorithm for finding the errors defined in (2.60) is developed. That algorithm seems to be computationally less involved than the algorithm presented in (2.65)-(2.67). The hybrid Newton/fixed-point iterations algorithm is given in Appendix 2.2.

2.7 Conclusion

In this chapter we have shown how to decompose an ill-defined singularly perturbed bilinear-quadratic optimal control problem into families of linear of linear-quadratic reduced-order well-defined control problems corresponding to the slow and fast subsystems. The results of this chapter can be also applied to the nonlinear singularly perturbed systems after they have been bilinearized (Schwarz et al., 1988). On the contrary to the linearization procedure (where all nonlinear terms are neglected), the system bilinearization preserves the nonlinear term representing the product of the state and control variables. Through this multiplicative term the control of bilinear systems is more effective than in the case of linear systems, where the control effects given system only through an additive term.

The results of this chapter can be extended to the problem of solving the bilinear-quadratic optimal control problems of singularly perturbed systems by the successive approximations technique to be presented in Chapter 4. This is left to the reader as an interesting research topic.

Appendix 2.1

Chang Decoupling Transformation and Its Version

Consider a linear time-invariant singularly perturbed system

$$\dot{x}_1 = A_1 x_1 + A_2 x_2 + B_1 u$$
$$\epsilon \dot{x}_2 = A_3 x_1 + A_4 x_2 + B_2 u \qquad\qquad (a.1)$$

Introducing the change of variables as

$$\eta_2 = x_2 + L x_1 \qquad\qquad (a.2)$$

the corresponding fast state equation in the new coordinates becomes

$$\epsilon \dot{\eta}_2 = (A_4 + \epsilon L A_2)\eta_2 + [A_3 + \epsilon L A_1 - (A_4 + \epsilon L A_2)L]x_1 + (B_2 + \epsilon L B_1)u \qquad\qquad (a.3)$$

By setting the coefficient multiplying x_1 in (a.3) to zero we get an independent pure-fast differential equation for η_2. This is possible since the obtained equation

$$A_3 + \epsilon L A_1 - (A_4 + \epsilon L A_2)L = 0 \qquad\qquad (a.4)$$

has a solution, which in the case when A_4 is nonsingular and ϵ is sufficiently small is unique.

Introducing another change of variables as

$$\eta_1 = x_1 - \epsilon H \eta_2 \qquad\qquad (a.5)$$

the slow differential equation from (a.1) becomes

$$\dot{\eta}_1 = (A_1 - A_2 L)\eta_1 + [A_2 + \epsilon(A_1 - A_2 L)H - H(A_4 + \epsilon L A_2)]\eta_2 + (B_1 - H B_2 - \epsilon H L B_1)u \qquad\qquad (a.6)$$

This equation becomes pure-slow if we set the coefficient multiplying η_2 to zero, that is

$$A_2 + \epsilon(A_1 - A_2 L)H - H(A_4 + \epsilon L A_2) = 0 \qquad\qquad (a.7)$$

which is feasible and implies the unique solution for H when A_4 is nonsingular and ϵ is sufficiently small. The celebrated Chang transformation (Chang, 1972) is defined by (a.2) and (a.5), that is

$$\begin{bmatrix} \eta_1 \\ \eta_2 \end{bmatrix} = \begin{bmatrix} I - \epsilon HL & -\epsilon H \\ L & I \end{bmatrix} \begin{bmatrix} x_1 \\ x_2 \end{bmatrix} = \mathbf{T}_1 \begin{bmatrix} x_1 \\ x_2 \end{bmatrix} \tag{a.8}$$

This nonsingular transformation has its inverse as

$$\mathbf{T}_1^{-1} = \begin{bmatrix} I & \epsilon H \\ -L & I - \epsilon LH \end{bmatrix} \tag{a.9}$$

Note that the same decomposition procedure is applicable in the case of time varying linear singularly perturbed systems. In that case the algebraic equations (a.4) and (a.7) are replaced by the differential ones with no initial conditions imposed on L and H.

In (Qureshi and Gajic, 1992) a new version of the Chang transformation is obtained by using the following change of variables

$$\begin{aligned} \eta_1 &= x_1 - \epsilon L_1 x_2 \\ \eta_2 &= -H_1 x_1 + x_2 \end{aligned} \tag{a.10}$$

In addition of decomposing the system equations into pure-slow and pure-fast, the transformation (a.10) also decomposes the transformation equations (a.4) and (a.7) so that they are independent of each other.

In the new coordinates, the system (a.1) under transformation (a.10) has the form

$$\begin{aligned} \dot{\eta}_1 &= (A_1 - L_1 A_3)\eta_1 \\ \epsilon \dot{\eta}_2 &= (A_4 - \epsilon H_1 A_2)\eta_2 \end{aligned} \tag{a.11}$$

while the transformation equations are given by

$$\begin{aligned} A_2 - L_1 A_4 + \epsilon(A_1 - L_1 A_3)L_1 &= 0 \\ A_3 + A_4 H_1 - \epsilon H_1(A_1 + A_2 H_1) &= 0 \end{aligned} \tag{a.12}$$

The time varying version of the new version of Chang transformation is also derived in (Qureshi and Gajic, 1992). In the time varying case the only difference is that the algebraic transformation equations become differential ones with no initial conditions imposed.

Appendix 2.2

Hybrid Newton Fixed-Point Iterations Algorithm for the Algebraic Riccati Equation of Singularly Perturbed Systems

In this appendix we present the hybrid Newton fixed-point iterations method for solving the algebraic Riccati equation of singularly perturbed systems by following the work of (Rutkowski and Gajic, 1993; Rutkowski, 1995). Consider

$$A^T P + PA + Q - PSP = 0 \tag{b.1}$$

with

$$A = \begin{bmatrix} A_1 & A_2 \\ \frac{1}{\epsilon}A_3 & \frac{1}{\epsilon}A_4 \end{bmatrix}, \quad S = \begin{bmatrix} S_1 & \frac{1}{\epsilon}S \\ \frac{1}{\epsilon}S^T & \frac{1}{\epsilon^2}S_3 \end{bmatrix}$$

$$Q = \begin{bmatrix} q_1^T q_1 & q_1^T q_2 \\ q_2^T q_1 & q_2^T q_2 \end{bmatrix}, \quad P = \begin{bmatrix} P_1 & \epsilon P_2 \\ \epsilon P_2^T & \epsilon P_3 \end{bmatrix} \tag{b.2}$$

where ϵ is a small positive singular perturbation parameter indicating the separation of the state variables into slow and fast ones.

It is known that the Newton method is convenient for the singularly perturbed systems whenever the first-order approximation (obtained by setting $\epsilon = 0$ in the partitioned equation (1)) is close to the exact one. In that case the Newton method is superior over the fixed-point type reduced-order parallel algorithm for solving this Riccati equation (Gajic and Shen, 1993), owing to its quadratic speed of convergence. In the following the solution of (b.1) will obtained by using the Newton method in terms of the reduced-order decoupled Lyapunov equations corresponding to the slow and fast subsystems.

Partitioning (b.1) according to (b.2) produces the following equations

$$P_1 A_1 + A_1^T P_1 + P_2 A_3 + A_3^T P_2^T - P_1 S P_2^T$$
$$- P_2 S^T P_1 - P_2 S_2 P_2^T + q_1^T q_1 = 0 \tag{b.3}$$

$$P_1 A_2 + P_2 A_4 + \epsilon A_1^T P_2 + A_3^T P_3 - \epsilon P_1 S_1 P_2$$
$$-P_1 S P_3 - \epsilon P_2 S^T P_2 - P_2 S_2 P_3 + q_1^T q_2 = 0 \tag{b.4}$$

$$P_3 A_4 + A_4^T P_3 + \epsilon P_2^T A_2 + \epsilon A_2^T P_2 - P_3 S_2 P_3$$
$$-\epsilon^2 P_2^T S_1 P_2 - \epsilon P_2^T S_1 P_2 - \epsilon P_3 S^T P_2 + q_2^T q_2 = 0 \tag{b.5}$$

where

$$S = B_1 R^{-1} B_2^T, \quad S_j = B_j R^{-1} B_j^T, \quad j = 1, 2 \tag{b.6}$$

The Newton method for solving nonlinear algebraic equations is based on the linearization procedure, namely, if we assume that $P_1^{(i)}, P_2^{(i)}, P_3^{(i)}$ are known points then the new iteration points are obtained as

$$P_j^{(i+1)} = P_j^{(i)} + \Delta P_j^{(i)}, \quad j = 1, 2, 3 \tag{b.7}$$

where $\Delta P_j^{(i)}$ are small quantities. Substituting (b.7) in (b.3)-(b.6) and neglecting $O(\Delta^2)$ terms as very small ones (linearization) we obtain the following set of equations

$$P_1^{(i+1)} D_1^{(i)} + D_1^{(i)^T} P_1^{(i+1)} + P_2^{(i+1)} D_3^{(i)} + D_3^{(i)^T} P_2^{(i+1)^T} + Q_1^{(i)} = 0 \tag{b.8}$$

$$P_1^{(i+1)} D_2^{(i)} + P_2^{(i+1)} D_4^{(i)} + D_3^{(i)^T} P_3^{(i+1)} + \epsilon D_1^{(i)^T} P_2^{(i+1)} + Q_2^{(i)} = 0 \tag{b.9}$$

$$P_3^{(i+1)} D_4^{(i)} + D_4^{(i)^T} P_3^{(i+1)} + \epsilon P_2^{(i+1)^T} D_2^{(i)} + \epsilon D_2^{(i)^T} P_2^{(i+1)} + Q_3^{(i)} = 0 \tag{b.10}$$

where

$$D_1^{(i)} = A_1 - S_1 P_1^{(i)} - S P_2^{(i)^T}, \quad D_2^{(i)} = A_2 - S P_3^{(i)} - \epsilon S_1 P_2^{(i)}$$

$$D_3^{(i)} = A_3 - S^T P_1^{(i)} - S_2 P_2^{(i)^T}, \quad D_4^{(i)} = A_4 - S_2 P_3^{(i)} - \epsilon S^T P_2^{(i)} \tag{b.11}$$

and

$$Q_1^{(i)} = P_1^{(i)} S_1 P_1^{(i)} + P_1^{(i)} S P_2^{(i)^T} + P_2^{(i)} S^T P_1^{(i)} + P_2^{(i)} S_2 P_2^{(i)^T} + q_1^T q_1$$

$$Q_2^{(i)} = \epsilon P_1^{(i)} S_1 P_2^{(i)} + P_1^{(i)} S P_3^{(i)} + \epsilon P_2^{(i)} S^T P_2^{(i)} + P_2^{(i)} S_2 P_3^{(i)} + q_1^T q_2$$

$$Q_3^{(i)} = P_3^{(i)} S_2 P_3^{(i)} + \epsilon^2 P_2^{(i)^T} S_1 P_2^{(i)} + \epsilon P_2^{(i)^T} S P_3^{(i)} + \epsilon P_3^{(i)} S^T P_2^{(i)} + q_2^T q_2$$
(b.12)

The optimal choice of the initial guesses $P_1^{(0)}, P_2^{(0)}, P_3^{(0)}$ for the proposed Newton method (b.8)-(b.10) will be considered later.

The application of the Newton method to the singularly perturbed partitioned algebraic Riccati equation leads to three linear coupled algebraic equations (b.8)-(b.10). These equations can be completely decoupled in the first approximation by setting $\epsilon = 0$ which produces

$$\mathbf{P}_1^{(i+1)} D_1^{(i)} + D_1^{(i)^T} \mathbf{P}_1^{(i+1)} + \mathbf{P}_2^{(i+1)} D_3^{(i)} + D_3^{(i)^T} \mathbf{P}_2^{(i+1)^T} + Q^{(i)} = 0 \quad \text{(b.13)}$$

$$\mathbf{P}_1^{(i+1)} D_2^{(i)} + \mathbf{P}_2^{(i+1)} D_4^{(i)} + D_3^{(i)^T} \mathbf{P}_3^{(i+1)} + Q_2^{(i)} = 0 \qquad \text{(b.14)}$$

$$\mathbf{P}_3^{(i+1)} D_4^{(i)} + D_4^{(i)^T} \mathbf{P}_3^{(i+1)} + Q_3^{(i)} = 0 \qquad \text{(b.15)}$$

From (b.14) we get

$$\mathbf{P}_2^{(i+1)} = -\left(\mathbf{P}_1^{(i+1)} D_2^{(i)} + D_3^{(i)^T} \mathbf{P}_3^{(i+1)} - Q_2^{(i)}\right) D_4^{(i)^{-1}} \qquad \text{(b.16)}$$

Substitution of (b.16) in (b.13) and elimination of $\mathbf{P}_3^{(i+1)}$ in the resulting equation by using equation (b.15) produces the reduced-order slow algebraic Lyapunov equation

$$\mathbf{P}_1^{(i+1)} D_0^{(i)} + D_0^{(i)^T} \mathbf{P}_1^{(i+1)} + Q_0^{(i)} = 0 \qquad \text{(b.17)}$$

where
$$D_0^{(i)} = D_1^{(i)} - D_2^{(i)} D_4^{(i)^{-1}} D_3^{(i)}$$

$$Q_0^{(i)} = Q_1^{(i)} + D_3^{(i)^T} D_4^{(i)^{-T}} Q_3^{(i)} D_4^{(i)^{-1}} D_3^{(i)} \qquad \text{(b.18)}$$

$$- Q_2^{(i)} D_4^{(i)^{-1}} D_3^{(i)} - D_3^{(i)^T} D_4^{(i)^{-T}} Q_2^{(i)^T}$$

Unique solutions to the first-order approximations (b.15)-(b.17) exist under the following assumption.

Assumption b.1 Matrices $D_0^{(i)}$ and $D_4^{(i)}$ are stable.

$$\triangle$$

This standard assumption in the theory of singular perturbations is always satisfied since $D_0^{(i)}$ and $D_4^{(i)}$ represent the closed-loop slow and fast subsystem matrices. Efficient numerical algorithms for solving the algebraic Lyapunov equations can be found in (Gajic and Qureshi, 1995).

In order to get solutions of (b.8)-(b.10) in terms of the solutions of (b.15)-(b.17) with any arbitrary order of accuracy, we define the error terms as

$$P_j^{(i+1)} = \mathbf{P}_j^{(i+1)} + \epsilon E_j \qquad \text{(b.19)}$$

Substituting (b.19) in (b.8)-(b.10) and using the fixed-point parallel algorithm from (Gajic and Shen, 1993) for solving the corresponding Lyapunov equations of the error terms, we get three completely decoupled linear equations

$$E_1^{(k+1)} D_0^{(i)} + D_0^{(i)^T} E_1^{(k+1)} = D_0^{(i)^T} \left(\mathbf{P}_2^{(i+1)} + \epsilon E_2^{(k)} \right) D_4^{(i)^{-1}} D_3^{(i)}$$
$$+ D_3^{(i)^T} D_4^{(i)^{-1}} \left(\mathbf{P}_2^{(i+1)} + \epsilon E_2^{(k)} \right)^T D_0^{(i)} \qquad \text{(b.20)}$$

$$E_3^{(k+1)} D_4^{(i)} + D_4^{(i)^T} E_3^{(k+1)} = -D_2^{(i)^T} \left(\mathbf{P}_2^{(i+1)} + \epsilon E_2^{(k)} \right)$$
$$- \left(\mathbf{P}_2^{(i+1)} + \epsilon E_2^{(k)} \right)^T D_2^{(i)} \qquad \text{(b.21)}$$

$$E_2^{(k+1)} = - \left(D_3^{(i)^T} E_3^{(k+1)} + E_1^{(k+1)} D_2^{(i)} + D_1^{(i)^T} \left(\mathbf{P}_2^{(i+1)} + \epsilon E_2^{(k)} \right) \right) D_4^{(i)^{-1}}$$
$$\text{(b.22)}$$

with starting point chosen as $E_1^{(0)} = 0$, $E_2^{(0)} = 0$, $E_3^{(0)} = 0$.

The fixed-point algorithm defined in (b.20)-(b.22) converges to the exact solution of the error terms with the rate of convergence of $O(\epsilon)$, that is (Gajic and Shen, 1993)

$$\left\| E_j - E_j^{(k)} \right\| = O\left(\epsilon^k\right), \quad k = 1, 2, \dots . \tag{b.23}$$

Using results from (b.23) in (b.19) implies that after k iterations we get

$$P_j^{(i+1)} = \mathbf{P}_j^{(i+1)} + \epsilon E_j^{(k)} + O\left(\epsilon^{k+1}\right) \tag{b.24}$$

After obtaining $P_j^{(i+1)}$ with the desired accuracy we go to the next iteration step with respect to i. Note that i iterations are the Newton method iterations so that one has to perform only 4–5 iterations with respect to i (the Newton method either converges in 4–5 iterations or does not converge at all).

Comparing the error equations (b.20)-(b.22) to the corresponding ones of the fixed-point algorithm for solving the algebraic Riccati equation of singularly perturbed systems (Gajic, 1986), it can be noticed that in the case of the Newton/fixed-point iterations algorithm the error equations have simpler form. The reason for that is the fact that in the case of the Newton/fixed-point algorithm we first perform linearization and then form the error equations. However, in the case of the fixed-point algorithm (Gajic, 1986) the error equations are obtained by doing algebra with the nonlinear (quadratic) terms, which causes the quite complex form for the error equations.

Very important factor for the convergence of the Newton method is the choice of good initial guesses. For the singularly perturbed systems we can use results from (Chow and Kokotovic, 1976), which state that

$$P_j = P_j^{(0)} + O(\epsilon), \quad j = 1, 2, 3 \tag{b.25}$$

where $P_j^{(0)}$ are given by following reduced-order slow and fast algebraic Riccati equations

$$P_1^{(0)}\mathbf{A} + \mathbf{A}^{\mathbf{T}}P_1^{(0)} - P_1^{(0)}\mathbf{S}P_1^{(0)} + \mathbf{Q} = 0 \tag{b.26}$$

$$P_3^{(0)} A_4 + A_4^T P_3^{(0)} - P_3^{(0)} S_2 P_3^{(0)} + q_2^T q_2 = 0 \qquad \text{(b.27)}$$

and

$$P_2^{(0)} = -\left(P_1^{(0)} A_2 + A_3^T P_3^{(0)} - P_1^{(0)} S P_3^{(0)} + q_1^T q_2 \right)\left(A_4 - S_2 P_3^{(0)} \right)^{-1}$$

$$\text{(b.28)}$$

with

$$\mathbf{A} = A_1 - A_2 A_4^{-1} A_3 - B_0 R_0^{-1} r^T Q_0$$

$$B_0 = B_1 - A_2 A_4^{-1} B_2, \quad Q_0 = q_1 - q_2 A_4^{-1} A_3$$
$$R_0 = R + r^T r, \quad r = -q_2 A_4^{-1} B_2 \qquad \text{(b.29)}$$

$$\mathbf{S} = B_0 R_0^{-1} B_0^T, \quad \mathbf{Q} = Q_0^T (I - r R_0^{-1} r^T) Q_0$$

Numerical examples that demonstrate the efficiency of the hybrid Newton fixed-point iterations algorithm can be found in (Rutkowski and Gajic, 1993; Rutkowski, 1995).

Chapter 3

Continuous-Time Weakly Coupled Bilinear Systems

3.1 Introduction

The purpose of this chapter is to study the optimal control problem of weakly coupled bilinear systems with a quadratic performance criterion. The weakly coupled bilinear systems have been studied so far only in the paper (Tzafestas and Anagnostou, 1984b), where the stabilization problem has been considered.

We study both the open-loop and approximate "closed-loop" optimal control problems of time invariant bilinear systems. The results are obtained in terms of the reduced-order problems (similarly to the singularly perturbed bilinear systems) for both the optimal open-loop and "closed-loop" control of weakly coupled bilinear systems. An extension of the presented methodology to the time varying weakly coupled bilinear systems can be done by paralleling the work presented in this chapter.

Many real physical systems are naturally weakly coupled such as power systems, communication satellites, helicopters, chemical reactors, electrical networks, flexible space structures, mechanical systems in modal coordinates.

The weakly linear coupled systems were introduced to the control audience by Kokotovic (Kokotovic et al., 1969). Since then many theoretical aspects for linear weakly coupled systems have been studied (Medanic and Avramovic, 1975; Ishimatsu et al., 1975; Ozguner and Perkins, 1977; Delacour at al., 1978; Mahmoud, 1978; Petkovski and Rakic, 1979; Washburn and Mendel, 1980; Arabacioglu et al., 1986; Petrovic and Gajic, 1988; Gajic and Shen, 1989; Harkara et al., 1989; Gajic et al., 1990; Shen and Gajic, 1990a, b, c; Su and Gajic, 1991, 1992; Skataric et al., 1993; Aganovic et al., 1994; Gajic and Aganovic, 1995).

The general weakly coupled systems, in different set-ups have been studied by Siljak, Basar and their coworkers (Ikeda and Siljak, 1980; Ohta and Siljak, 1985; Sezer and Siljak, 1986, 1991; Kaszkurewicz et al., 1990; Siljak, 1991; Srikant and Basar, 1989, 1991, 1992a, b). The weak coupling has been also considered in the concept of multimodeling (Khalil and Kokotovic, 1978; Ozguner, 1979; Khalil, 1980; Saksena and Cruz, 1981a, b; Saksena and Basar, 1982; Gajic and Khalil, 1986; Gajic, 1988, Zhuang and Gajic, 1991), jump parameter linear optimal control systems (Borno and Gajic, 1995), and for nearly completely decomposable Markov chains (Delebecque and Quadrat, 1981; Srikant and Basar, 1989; Aldhaheri and Khalil, 1991). The nonlinear weakly coupled systems have been only studied in a few papers (Kokotovic and Singh, 1971; Srikant and Basar, 1991, 1992).

3.2 Optimal Control of Weakly Coupled Bilinear Systems

A sequence of linear state and costate equations is constructed such that the open-loop solution of the bilinear-quadratic dynamic optimization problem is obtained in terms of the reduced-order subsystems. The obtained results utilize the recursive scheme for the optimal control of a general bilinear system with a quadratic performance criterion (Hofer and Tibken, 1988) and the time varying version of the reduced-order method for solving the linear-quadratic optimal open-loop weakly coupled control problem (Su and Gajic, 1991). This leads to the reduction in the size of the required computation and allows parallel processing of information.

The near-optimal "closed-loop" control is obtained in the form of an approximate linear "feedback" law, with the feedback gains calculated from two reduced-order independent time varying linear-quadratic optimal control problems. The obtained results are based on the idea of the recursive reduced-order scheme for solving the algebraic Riccati equation of weakly coupled systems (Gajic et al., 1990) and the recursive scheme for the optimal control of general bilinear systems with quadratic performance criteria (Cebuhar and Constanza, 1984). An algorithm which produces an arbitrary degree of accuracy for the "closed-loop" control is derived. The results are demonstrated on a real physical bilinear model of a paper making machine.

3.2.1 Open-Loop Control of Weakly Coupled Bilinear Systems

In this section, we exploit the iterative scheme presented in Section 2.3 comprising a sequence of linear two-point boundary value problems, in order to derive the solution for the optimal open-loop control of weakly coupled bilinear systems. The solution is obtained in the spirit of the general theory of small parameter control problems, namely, the problem is decomposed into two reduced-order subproblems. The open-loop optimal control of weakly coupled linear systems has been studied in (Su and Gajic, 1991). The study of (Su and Gajic, 1991) is limited to the time invariant systems. In this section, we show that following the ideas of (Su and Gajic, 1991) we are able to handle in the same manner the time varying weakly coupled two-point boundary value problem.

The weakly coupled bilinear control system under consideration is represented by

$$
\begin{bmatrix} \dot{y}_1 \\ \dot{y}_2 \end{bmatrix} = \begin{bmatrix} A_1 & \epsilon A_2 \\ \epsilon A_3 & A_4 \end{bmatrix} \begin{bmatrix} y_1 \\ y_2 \end{bmatrix} + \begin{bmatrix} B_1 & \epsilon B_2 \\ \epsilon B_3 & B_4 \end{bmatrix} \begin{bmatrix} u_1 \\ u_2 \end{bmatrix}
$$

$$
+ \left\{ \begin{bmatrix} y_1 \\ y_2 \end{bmatrix} \begin{bmatrix} M_a & \epsilon M_b \\ \epsilon M_c & M_d \end{bmatrix} \right\} \begin{bmatrix} u_1 \\ u_2 \end{bmatrix}, \qquad \begin{bmatrix} y_1(t_0) \\ y_2(t_0) \end{bmatrix} = \begin{bmatrix} y_1^0 \\ y_2^0 \end{bmatrix}
$$

(3.1)

where $y_1 \in \Re^{n_1}$, $y_2 \in \Re^{n_2}$, $u_i \in \Re^{m_i}$, $i = 1, 2$, and ϵ is a small coupling parameter, with

$$\left\{ \begin{bmatrix} y_1 \\ y_2 \end{bmatrix} \begin{bmatrix} M_1 & \epsilon M_2 \\ \epsilon M_3 & M_4 \end{bmatrix} \right\} = \sum_{i=1}^{n_1} y_{1i} \begin{bmatrix} M_{ai} & M_{bi} \\ M_{ci} & M_{di} \end{bmatrix}$$

$$+ \sum_{j=n_1+1}^{n_1+n_2} y_{2(j-n_1)} \begin{bmatrix} M_{aj} & M_{bj} \\ M_{cj} & M_{dj} \end{bmatrix}$$

(3.2)

where $M_{ai} \in \Re^{n_1 \times m_1}$, $M_{bi} \in \Re^{n_1 \times m_2}$, $M_{ci} \in \Re^{n_2 \times m_1}$, $M_{di} \in \Re^{n_2 \times m_2}$.
A quadratic cost functional to be minimized is associated with (3.1) having the following form

$$J = \frac{1}{2} \int_{t_0}^{t_f} \left(\begin{bmatrix} y_1 \\ y_2 \end{bmatrix}^T Q \begin{bmatrix} y_1 \\ y_2 \end{bmatrix} + \begin{bmatrix} u_1 \\ u_2 \end{bmatrix}^T R \begin{bmatrix} u_1 \\ u_2 \end{bmatrix} \right) dt$$

$$+ \frac{1}{2} \begin{bmatrix} y_1(t_f) \\ y_2(t_f) \end{bmatrix}^T F \begin{bmatrix} y_1(t_f) \\ y_2(t_f) \end{bmatrix}$$

(3.3)

with $Q \geq 0$, $R > 0$, $F \geq 0$ possessing the weak coupling structures, that is

$$Q = \begin{bmatrix} Q_1 & \epsilon Q_2 \\ \epsilon Q_2^T & Q_3 \end{bmatrix}, \quad R = \begin{bmatrix} R_1 & 0 \\ 0 & R_2 \end{bmatrix}, \quad F = \begin{bmatrix} F_1 & \epsilon F_2 \\ \epsilon F_2^T & F_3 \end{bmatrix}$$

(3.4)

In the following, we will utilize the recursive scheme (2.9)-(2.14) in order to find the optimal open-loop control law for the weakly coupled bilinear-quadratic optimal control problem represented by (3.1)-(3.4) in terms of the reduced-order subsystems.

It can be shown, after some algebra, that the system of equations (2.9) preserves the weak coupling structure. Namely, the use of (3.1)-(3.4) in (2.4)-(2.5) and (2.9)-(2.11) produces

$$\begin{bmatrix} \dot{y}_1 \\ \dot{y}_2 \end{bmatrix}^{(k+1)} = \tilde{A}^{(k)} \begin{bmatrix} y_1 \\ y_2 \end{bmatrix}^{(k+1)} - \tilde{B}^{(k)} R^{-1} \tilde{B}^{(k)T} \begin{bmatrix} q_1 \\ q_2 \end{bmatrix}^{(k+1)}$$

$$\begin{bmatrix} \dot{q}_1 \\ \dot{q}_2 \end{bmatrix}^{(k+1)} = -\tilde{Q}^{(k)} \begin{bmatrix} y_1 \\ y_2 \end{bmatrix}^{(k+1)} - \tilde{A}^{(k)T} \begin{bmatrix} q_1 \\ q_2 \end{bmatrix}^{(k+1)}$$

(3.5)

where $q_1 \in \Re^{n_1}$ and $q_2 \in \Re^{n_2}$ are costate vectors corresponding, respectively, to the state variables y_1 and y_2. The time varying matrices in (3.5) are given by

$$\widetilde{A}^{(k)} = \begin{bmatrix} \widetilde{A}_1^{(k)} & \epsilon\widetilde{A}_2^{(k)} \\ \epsilon\widetilde{A}_3^{(k)} & \widetilde{A}_4^{(k)} \end{bmatrix}, \qquad \widetilde{Q}^{(k)} = \begin{bmatrix} \widetilde{Q}_1^{(k)} & \epsilon\widetilde{Q}_2^{(k)} \\ \epsilon\widetilde{Q}_2^{(k)T} & \widetilde{Q}_3^{(k)} \end{bmatrix}$$

$$\widetilde{S}^{(k)} = \widetilde{B}^{(k)}R^{-1}\widetilde{B}^{(k)T} = \begin{bmatrix} \widetilde{S}_1^{(k)} & \epsilon\widetilde{S}_2^{(k)} \\ \epsilon\widetilde{S}_2^{(k)T} & \widetilde{S}_3^{(k)} \end{bmatrix} \tag{3.6}$$

Note that partitions defined in (3.6) have to be performed by a computer only, in the process of computations, and there is no need for the corresponding analytical expressions. After some algebra the state-costate equations (3.5) can be written in the form

$$\begin{bmatrix} \dot{y}_1^{(k+1)} \\ \dot{q}_1^{(k+1)} \\ \dot{y}_2^{(k+1)} \\ \dot{q}_2^{(k+1)} \end{bmatrix} = \begin{bmatrix} \widetilde{T}_1^{(k)} & \epsilon\widetilde{T}_2^{(k)} \\ \epsilon\widetilde{T}_3^{(k)} & \widetilde{T}_4^{(k)} \end{bmatrix} \begin{bmatrix} y_1^{(k+1)} \\ q_1^{(k+1)} \\ y_2^{(k+1)} \\ q_2^{(k+1)} \end{bmatrix} \tag{3.7}$$

The time varying matrices $\widetilde{T}_i^{(k)}$ introduced in (3.7) have the following structures

$$\widetilde{T}_1^{(k)} = \begin{bmatrix} \widetilde{A}_1^{(k)} & -\widetilde{S}_1^{(k)} \\ -\widetilde{Q}_1^{(k)} & -\widetilde{A}_1^{(k)T} \end{bmatrix}, \qquad \widetilde{T}_2^{(k)} = \begin{bmatrix} \widetilde{A}_2^{(k)} & -\widetilde{S}_2^{(k)} \\ -\widetilde{Q}_2^{(k)} & -\widetilde{A}_3^{(k)T} \end{bmatrix}$$

$$\widetilde{T}_3^{(k)} = \begin{bmatrix} \widetilde{A}_3^{(k)} & -\widetilde{S}_2^{(k)T} \\ -\widetilde{Q}_2^{(k)T} & -\widetilde{A}_2^{(k)T} \end{bmatrix}, \qquad \widetilde{T}_4^{(k)} = \begin{bmatrix} \widetilde{A}_4^{(k)} & -\widetilde{S}_2^{(k)} \\ -\widetilde{Q}_3^{(k)} & -\widetilde{A}_4^{(k)T} \end{bmatrix} \tag{3.8}$$

The expression for the boundary conditions is changed due to an interchange of rows corresponding to $q_1^{(k+1)}$ and $y_2^{(k+1)}$, which modifies matrices defined in (2.12) as follows

$$U_1 \begin{bmatrix} w^{(k+1)}(t_0) \\ \lambda^{(k+1)}(t_0) \end{bmatrix} + V_1 \begin{bmatrix} w^{(k+1)}(t_f) \\ \lambda^{(k+1)}(t_f) \end{bmatrix} = c_1 \tag{3.9}$$

with new notation

$$\begin{bmatrix} y_1^{(k+1)} \\ q_1^{(k+1)} \end{bmatrix} = w^{(k+1)}, \qquad \begin{bmatrix} y_2^{(k+1)} \\ q_2^{(k+1)} \end{bmatrix} = \lambda^{(k+1)} \tag{3.10}$$

and

$$U_1 = \begin{bmatrix} I_{n_1} & 0 & 0 & 0 \\ 0 & 0 & 0 & 0 \\ 0 & 0 & I_{n_2} & 0 \\ 0 & 0 & 0 & 0 \end{bmatrix}, \quad c_1 = \begin{bmatrix} y_1^0 \\ 0 \\ y_2^0 \\ 0 \end{bmatrix}$$

(3.11)

$$V_1 = \begin{bmatrix} 0 & 0 & 0 & 0 \\ -F_1 & I_{n_1} & -\epsilon F_2 & 0 \\ 0 & 0 & 0 & 0 \\ -F_2^T & 0 & -F_3 & I_{n_2} \end{bmatrix}$$

In order to obtain the decoupled subsystems from (3.7), we apply the transformation of (Gajic and Shen, 1989) given by

$$\begin{bmatrix} \eta^{(k+1)} \\ \xi^{(k+1)} \end{bmatrix} = \mathbf{T}_2^{(k)}(t, \epsilon) \begin{bmatrix} w^{(k+1)} \\ \lambda^{(k+1)} \end{bmatrix}$$

(3.12)

with

$$\mathbf{T}_2^{(k)}(t, \epsilon) = \begin{bmatrix} I_1 & -\epsilon L^{(k)} \\ \epsilon H^{(k)} & I_2 - \epsilon^2 H^{(k)} L^{(k)} \end{bmatrix}$$

(3.13)

$$\mathbf{T}_2^{(k)-1}(t, \epsilon) = \begin{bmatrix} I_1 - \epsilon^2 H^{(k)} L^{(k)} & \epsilon L^{(k)} \\ -\epsilon H^{(k)} & I_2 \end{bmatrix}$$

where I_1 and I_2 are identity matrices of order $2n_1$ and $2n_2$, respectively. The matrices $L^{(k)}$ and $H^{(k)}$ are obtained as the solutions of the following nonlinear differential equations

$$\dot{L}^{(k)} = \tilde{T}_1^{(k)} L^{(k)} - L^{(k)} \tilde{T}_4^{(k)} + \tilde{T}_2^{(k)} - \epsilon^2 L^{(k)} \tilde{T}_3^{(k)} L^{(k)}$$

$$\dot{H}^{(k)} = H^{(k)} \left(\tilde{T}_1^{(k)} - \epsilon^2 L^{(k)} \tilde{T}_3^{(k)} \right) - \left(\tilde{T}_4^{(k)} + \epsilon^2 \tilde{T}_3^{(k)} L^{(k)} \right) H^{(k)} + \tilde{T}_3^{(k)}$$

(3.14)

The initial conditions for differential equations (3.14) are arbitrary (Qureshi and Gajic, 1991). The existence of the bounded solutions of (3.82) for sufficiently small values ϵ is established in (Qureshi and Gajic, 1991; Qureshi, 1992).

The transformation (3.12) applied to the system (3.7) produces two completely decoupled subsystems

$$\dot{\eta}^{(k+1)} = \left(\tilde{T}_1^{(k)} - \epsilon^2 L^{(k)} \tilde{T}_3^{(k)}\right)\eta^{(k+1)} \tag{3.15}$$

$$\dot{\xi}^{(k+1)} = \left(\tilde{T}_4^{(k)} - \epsilon^2 \tilde{T}_3^{(k)} L^{(k)}\right)\xi^{(k+1)} \tag{3.16}$$

Consequently, the change of variables transforms the boundary conditions into

$$U_2 \begin{bmatrix} \eta^{(k+1)}(t_0) \\ \xi^{(k+1)}(t_0) \end{bmatrix} + V_2 \begin{bmatrix} \eta^{(k+1)}(t_f) \\ \xi^{(k+1)}(t_f) \end{bmatrix} = c_1 \tag{3.17}$$

where

$$U_2 = U_1 \mathbf{T}_2^{(k)^{-1}}(t_0, \epsilon), \quad . \quad V_2 = V_1 \mathbf{T}_2^{(k)^{-1}}(t_f, \epsilon) \tag{3.18}$$

The solutions of linear systems of time varying differential equations (3.15) and (3.16) are

$$\eta^{(k+1)}(t) = \Phi^{(k)}(t, t_0, \epsilon)\eta^{(k+1)}(t_0)$$

$$\xi^{(k+1)}(t) = \Psi^{(k)}(t, t_0, \epsilon)\xi^{(k+1)}(t_0) \tag{3.19}$$

where $\Phi^{(k)}(t, t_0, \epsilon)$ and $\Psi^{(k)}(t, t_0, \epsilon)$ are the transition matrices of (3.15) and (3.16), respectively. It is assumed that $\Phi^{(k)}(t, t_0, \epsilon)$ and $\Psi^{(k)}(t, t_0, \epsilon)$ are known for every t. The initial conditions $\eta^{(k+1)}(t_0)$ and $\xi^{(k+1)}(t_0)$ have to be determined. This can be done as follows. Substitution of (3.19) into (3.17) yields

$$\Delta^{(k+1)}(\epsilon) \begin{bmatrix} \eta^{(k+1)}(t_0) \\ \xi^{(k+1)}(t_0) \end{bmatrix} = c_1 \tag{3.20}$$

where

$$\Delta^{(k)}(\epsilon) = U_2(\epsilon) + V_2(\epsilon) \begin{bmatrix} \Phi^{(k)}(t_f, t_0, \epsilon) & 0 \\ 0 & \Psi^{(k)}(t_f, t_0, \epsilon) \end{bmatrix} \tag{3.21}$$

If $\Delta^{(k+1)^{-1}}(\epsilon)$ exists then the solution of (3.20) will be given by

$$\begin{bmatrix} \eta^{(k+1)}(t_0) \\ \xi^{(k+1)}(t_0) \end{bmatrix} = \Delta^{(k+1)^{-1}}(\epsilon) c_1 \tag{3.22}$$

Note that as $\epsilon \to 0$

$$\left\{ \mathbf{T}_2^{(k)}(t,0) \right\}^{-1} = \begin{bmatrix} I_1 & 0 \\ 0 & I_2 \end{bmatrix} = I \tag{3.23}$$

The transition matrices $\Phi^{(k)}(t,t_0,0)$ and $\Psi^{(k)}(t,t_0,0)$ can be partitioned in blocks of the same dimensions

$$\Phi^{(k)}(t,t_0,0) = \begin{bmatrix} \Phi_{11}^{(k)}(t,t_0,0) & \Phi_{12}^{(k)}(t,t_0,0) \\ \Phi_{21}^{(k)}(t,t_0,0) & \Phi_{22}^{(k)}(t,t_0,0) \end{bmatrix}$$

$$\Psi^{(k)}(t,t_0,0) = \begin{bmatrix} \Psi_{11}^{(k)}(t,t_0,0) & \Psi_{12}^{(k)}(t,t_0,0) \\ \Psi_{21}^{(k)}(t,t_0,0) & \Psi_{22}^{(k)}(t,t_0,0) \end{bmatrix}$$

After doing some algebra, it can be shown the matrix defined in (3.21) is given by

$$\Delta^{(k+1)}(\epsilon) = \begin{bmatrix} I_{n_1} & 0 & 0 & 0 \\ 0 & \Delta_{22}^{(k+1)}(0) & 0 & 0 \\ 0 & 0 & I_{n_2} & 0 \\ 0 & 0 & 0 & \Delta_{44}^{(k+1)}(0) \end{bmatrix} + O(\epsilon) \tag{3.24}$$

where

$$\Delta_{22}^{(k+1)}(0) = \Phi_{22}^{(k)}(t_f,t_0,0) - F_1 \Phi_{12}^{(k)}(t_f,t_0,0)$$

$$\tag{3.25}$$

$$\Delta_{44}^{(k+1)}(0) = \Psi_{22}^{(k)}(t_f,t_0,0) - F_3 \Psi_{12}^{(k)}(t_f,t_0,0)$$

Since both matrices $\Phi_{22}^{(k)}(t_f,t_0,0) - F_1 \Phi_{12}^{(k)}(t_f,t_0,0)$ and $\Psi_{22}^{(k)}(t_f,t_0,0) - F_3 \Psi_{12}^{(k)}(t_f,t_0,0)$ are nonsingular for all choices of F_1 and F_3 (Kalman, 1960; see also Kirk, 1970, page 211), so does $\Delta^{(k+1)}(\epsilon)$ for $0 < \epsilon \le \epsilon_1$ and ϵ_1 sufficiently small. Thus, in summary, we have established the following theorem.

Theorem 3.1 *Let the problem matrices be continuous functions of t on the time interval $t_0 \le t \le t_f$, then for all sufficiently small ϵ the boundary value problem (3.15)–(3.18) has the solution given by*

$$\begin{bmatrix} \eta^{(k+1)}(t,\epsilon) \\ \xi^{(k+1)}(t,\epsilon) \end{bmatrix} = \begin{bmatrix} \Phi^{(k+1)}(t,t_0,\epsilon) & 0 \\ 0 & \Psi^{(k+1)}(t,t_0,\epsilon) \end{bmatrix} \Delta^{(k+1)^{-1}}(\epsilon) c_1$$

*Consequently, the solution of the original boundary problem (3.7)-(3.11) is
obtained from (3.12) as*

$$\begin{bmatrix} w^{(k+1)}(t,\epsilon) \\ \lambda^{(k+1)}(t,\epsilon) \end{bmatrix} = \{\mathbf{T_2}(t,\epsilon)\}^{-1} \begin{bmatrix} \eta^{(k+1)}(t,\epsilon) \\ \xi^{(k+1)}(t,\epsilon) \end{bmatrix} \tag{3.26}$$

*so that the required variables $y_1^{(k+1)}$ and $y_2^{(k+1)}$ are obtained by partitioning
the vectors $w^{(k+1)}$ and $\lambda^{(k+1)}$ according to (3.10). The same holds for the
costate variables, $q_1^{(k+1)}$ and $q_2^{(k+1)}$, that is, they are obtained form (3.12)
and (3.10).*

∎

Having obtained the approximate state trajectories $y_1^{(k+1)}$ and $y_2^{(k+1)}$
and the approximate costate trajectories $q_1^{(k+1)}$ and $q_2^{(k+1)}$, the approximate
optimal open-loop control can be expressed as

$$u^{(k+1)}(t) = -R^{-1}\left(B + \left\{\begin{bmatrix} y_1^{(k+1)}(t) \\ y_2^{(k+1)}(t) \end{bmatrix} M\right\}\right)^T \begin{bmatrix} q_1^{(k+1)}(t) \\ q_2^{(k+1)}(t) \end{bmatrix} \tag{3.27}$$

The main problem that we are faced with in the presented method is the
problem of finding the transition matrices $\Phi^{(k)}(t,t_0,\epsilon)$ and $\Psi^{(k)}(t,t_0,\epsilon)$ of the
corresponding linear time varying systems. This can be done, in general, only
numerically. Another way to overcome this problem is to study the optimal
open-loop control of weakly coupled system in the discrete-time domain,
where the system transition matrices can be presented in the analytical form.
Research in that direction is underway.

3.2.2 "Closed-Loop" Control of Weakly Coupled Bilinear Systems

In this section we present an approximate solution to the bilinear-quadratic
optimal control problem, which has the nature of the closed-loop solution.
This has been done by following the same methodology like in the case
of singularly perturbed systems considered in Section 2.4. For the weakly
coupled sequence of linear systems represented by (2.44), which approximate
the solution of the bilinear-quadratic optimal control problem (3.1)-(3.4), the

matrices $B_i(t), S_i(t), P_i(t)$ in equations (2.44)-(2.45) can be partitioned as

$$B_i(t) = \begin{bmatrix} B_{1i}(t) & \epsilon B_{2i}(t) \\ \epsilon B_{3i}(t) & B_{4i}(t) \end{bmatrix}, \quad S_i(t) = \begin{bmatrix} S_{1i}(t) & \epsilon S_{2i}(t) \\ \epsilon S_{2i}^T(t) & S_{3i} \end{bmatrix}$$

$$P_i(t) = \begin{bmatrix} P_{1i}(t) & \epsilon P_{2i}(t) \\ \epsilon P_{2i}^T(t) & P_{3i}(t) \end{bmatrix}$$

(3.28)

Partitioning the algebraic Riccati equation given in (2.45) according to (3.28) and setting $\epsilon = 0$, we get an $O(\epsilon^2)$ approximation of (2.45) in terms of the reduced-order, decoupled algebraic Riccati equations

$$\mathbf{P_{1i}}(t)A_1 + A_1^T\mathbf{P_{1i}}(t) + Q_1 - \mathbf{P_{1i}}(t)S_{1i}(t)\mathbf{P_{1i}}(t) = 0$$

$$\mathbf{P_{3i}}(t)A_4 + A_4^T\mathbf{P_{3i}}(t) + Q_3 - \mathbf{P_{3i}}(t)S_{3i}(t)\mathbf{P_{3i}}(t) = 0$$

(3.29)

$$\mathbf{P_{2i}}(t)(A_4 - S_{3i}(t)\mathbf{P_{3i}}(t)) + (A_1 - S_{1i}(t)\mathbf{P_{1i}}(t))^T\mathbf{P_{2i}}(t)$$
$$+ \mathbf{P_{1i}}(t)A_2 + A_2^T\mathbf{P_{3i}}(t) + Q_2 - \mathbf{P_{1i}}(t)S_{si}(t)\mathbf{P_{3i}}(t) = 0$$

The unique positive semidefinite solution of (3.29) exists under the following assumption.

Assumption 3.1 The triples $(A_1, B_{1i}(t), \sqrt{Q_1})$ and $(A_4, B_{4i}(t), \sqrt{Q_3})$ are stabilizable-detectable for every t.

△

Corresponding solution, $\mathbf{P_i}(t)$, defined as

$$\mathbf{P_i}(t) = \begin{bmatrix} \mathbf{P_{1i}}(t) & \epsilon\mathbf{P_{2i}}(t) \\ \epsilon\mathbf{P_{2i}^T}(t) & \mathbf{P_{3i}}(t) \end{bmatrix}$$

(3.30)

is $O(\epsilon^2)$ close to the optimal one, $P_i(t)$. An $O(\epsilon^2)$ perturbation made in the iterative scheme (2.44)-(2.46) propagates into next iteration, but due to the continuous dependence of the solution of the sequence of linear differential equations with respect to a perturbation in the system coefficients, the presented method produces

$$x_i(t) = \mathbf{x_i}(t) + O(\epsilon^2), \quad i = 1, 2, ..., \quad \forall t \geq 0$$

(3.31)

and

$$u_i^{app}(t) = \mathbf{u_i^{app}}(t) + O(\epsilon^2), \quad i = 1, 2, ..., \quad \forall t \geq 0 \qquad (3.32)$$

where

$$\dot{\mathbf{x}}_i(t) = A\mathbf{x}_i(t) + \mathbf{B}_i(t)\mathbf{u}_i^{app}(t) \qquad (3.33)$$

$$\mathbf{u}_i^{app}(t) = -R^{-1}\mathbf{B}_i(t)\mathbf{P}_i(t)\mathbf{x}_i(t) \qquad (3.34)$$

with

$$\mathbf{B}_i(t) = B + \{\mathbf{x}_{i-1}(t)M\} \qquad (3.35)$$

If one intends to improve the accuracy of the solution of the Riccati equation (2.45), one can use (in the last iteration with respect to i only) an iterative refinement of (Gajic et al., 1990; Shen and Gajic, 1990a). Define the approximations of the required solution of (2.45) as

$$P_{ji}^{(k)}(t, \epsilon) = \mathbf{P}_{ji}(t, \epsilon) + \epsilon^2 E_{ji}^{(k)}(t, \epsilon) \quad j = 1, 2, 3 \qquad (3.36)$$

Then, the recursive reduced-order scheme for the error equations are obtained in therms of the reduced-order algebraic Lyapunov equations (Gajic et al., 1990)

$$E_{1i}^{(k+1)}\Delta_{1i} + \Delta_{1i}^T E_{1i}^{(k+1)} = M_{1i}^{(k)}$$

$$E_{3i}^{(k+1)}\Delta_{4i} + \Delta_{4i}^T E_{3i}^{(k+1)} = M_{3i}^{(k)} \qquad (3.37)$$

$$E_{2i}^{(k+1)}\Delta_{4i} + \Delta_{1i}^T E_{2i}^{(k+1)} + E_{1i}^{(k+1)}\Delta_{2i} + \Delta_{3i}^T E_{3i}^{(k+1)} = M_{2i}^{(k,k+1)}$$

for $k = 0, 1, 2, ...$, with the initial conditions given by

$$E_{1i}^{(0)} = 0, \ E_{2i}^{(0)} = 0, \ E_{3i}^{(0)} = 0$$

Methods for solving the algebraic Lyapunov equations are summarized in (Gajic and Qureshi, 1995). The matrices Δ_{ij}, $j = 1, ..., 4$ and

$M_{1i}^{(k)}, M_{2i}^{(k,k+1)}, M_{3i}^{(k)}$ are given by

$$\Delta_{1i} = A_1 - S_{1i}P_{1i}, \quad \Delta_{2i} = A_2 - S_{1i}P_{2i} - S_2 P_{3i}$$

(3.38)

$$\Delta_{4i} = A_4 - S_{3i}P_{3i}, \quad \Delta_{3i} = A_3 - S_{3i}P_{2i}^T - S_2^T P_{1i}$$

and

$$M_{1i}^{(k)} = P_{2i}^{(k)} S_{2i}^T P_{1i}^{(k)} + P_{1i}^{(k)} S_{2i} P_{2i}^{(k)^T} + P_{2i}^{(k)} S_{3i} P_{2i}^{(k)^T}$$
$$- P_{2i}^{(k)} A_3 - A_3^T P_{2i}^{(k)^T} - \epsilon^2 E_{1i}^{(k)} S_{1i} E_{1i}^{(k)}$$

(3.39)

$$M_{3i}^{(k)} = P_{3i}^{(k)} S_{2i}^T P_{2i}^{(k)} + P_{2i}^{(k)^T} S_{2i} P_{3i}^{(k)} + P_{2i}^{(k)^T} S_{1i} P_{2i}^{(k)}$$
$$+ P_{3i}^{(k)} S_{2i}^T P_{3i}^{(k)} - P_{2i}^{(k)^T} A_2 - A_2^T P_{2i}^{(k)} + \epsilon^2 E_{3i}^{(k)} S_{3i} E_{3i}^{(k)}$$

$$M_{2i}^{(k,k+1)} = P_{2i}^{(k)} S_{2i}^T P_{2i}^{(k)} + \epsilon^2 E_{1i}^{(k+1)} S_{1i} E_{2i}^{(k)} + \epsilon^2 E_{2i}^{(k)} S_{3i} E_{3i}^{(k+1)}$$
$$+ \epsilon^2 E_{1i}^{(k+1)} S_{2i} E_{3i}^{(k+1)}$$

Note that under Assumption 3.1 both Δ_{1i} and Δ_{4i} are stable matrices.

It can be shown that the rate of convergence of (3.36)-(3.39) is $O(\epsilon^2)$ (Gajic et al., 1990), that is

$$\left\| E_{ji}^{(k+1)} - E_{ji}^{(k)} \right\| = O(\epsilon^2), \quad i = 0, 1, 2, \ldots; \ j = 1, 2, 3; \ k = 0, 1, \ldots \quad (3.40)$$

which implies

$$\left\| P_{ji} - P_{ji}^{(k)} \right\| = O\left(\epsilon^{2(k+1)}\right), \quad i = 0, 1, 2, \ldots; \ j = 1, 2, 3; \ k = 0, 1, \ldots$$

(3.41)

Having obtained $P_i(t)$ with the accuracy of $O\left(\epsilon^{2(k+1)}\right)$ produces the same accuracy for the approximations of the optimal state trajectories and optimal control laws.

Note that the above improvement algorithm (3.37)-(3.41) can be also implemented by using the Newton method instead the fixed point iterations for solving the corresponding algebraic weakly coupled Riccati equation (see Appendix 3.2).

Results from this section are mostly based on the work from (Aganovic, 1993; Aganovic and Gajic, 1993).

3.3 Case Study: A Paper Making Machine

In order to demonstrate the efficiency of the proposed method for the near-optimal "closed-loop" control of weakly coupled bilinear systems we have run a fourth-order real world example, a paper making machine control problem (Ying et al., 1992). The bilinear mathematical model of this system is formulated according to (3.1) and (3.3) as

$$A = \begin{bmatrix} -1.93 & 0 & 0 & 0 \\ 0.394 & -0.426 & 0 & 0 \\ 0 & 0 & -0.63 & 0 \\ 0.095 & -0.103 & 0.413 & -0.426 \end{bmatrix}, \quad B = \begin{bmatrix} 1.274 & 1.274 \\ 0 & 0 \\ 1.34 & -0.65 \\ 0 & 0 \end{bmatrix}$$

$$N_1 = \begin{bmatrix} 0 & 0 \\ 0 & 0 \\ 0.755 & 0.366 \\ 0 & 0 \end{bmatrix}, N_2 = N_4 = \begin{bmatrix} 0 & 0 \\ 0 & 0 \\ 0 & 0 \\ 0 & 0 \end{bmatrix}, N_3 = \begin{bmatrix} 0 & 0 \\ 0 & 0 \\ -0.718 & -0.718 \\ 0 & 0 \end{bmatrix}$$

Weighting matrices Q and R chosen as

$$Q = \begin{bmatrix} 1 & 0 & 0.13 & 0 \\ 0 & 1 & 0 & 0.09 \\ 0.13 & 0 & 0.1 & 0 \\ 0 & 0.09 & 0 & 0.2 \end{bmatrix}, \quad R = \begin{bmatrix} 1 & 0 \\ 0 & 1 \end{bmatrix}$$

Note that the matrices B, N_1, and N_2 have no weakly coupled forms. However, it has been shown in (Skataric et al., 1991) that the classes of linear-quadratic optimal control problems having weakly coupled system matrix and strongly coupled input matrix can be studied as the weakly coupled linear-quadratic optimal control problems by assuming the special form for the state penalty matrix. Small perturbation parameter is equal to $\epsilon = 0.1$. Simulation results, obtained by using the MATLAB package, are presented in Figures 3.1–3.6. Figures 3.1–3.6 represent the approximate and optimal trajectories and the approximate and the optimal controls. The optimal ones are represented by the solid lines. It can be seen from the these plots that the approximate trajectories and controls are very good approximations for the optimal ones.

The number of iterations performed are $i = 3$ and $k = 1$, where i represents the number of linear time varying systems in the sequence defined by (2.44) and k represents the number of iterations performed to increase the accuracy as defined by (3.36).

3.4 Conclusion

It has been shown how to exploit the presence of a small weak coupling parameter and get the near-optimal controllers for interconnected (weakly coupled) systems in terms of the reduced-order subsystems. We have limited our attention to the systems composed of only two subsystems. It is interesting to expend the presented methodology to the linear weakly coupled systems composed of many subsystems.

The results of this chapter can be applied to the nonlinear weakly coupled systems after they have been bilinearized. It should be point out that mathematical models of many mechanical systems are nonlinear and weakly coupled. Even more, in the case of mechanical systems described by partial differential equations and presented in the modal coordinates (Meirovich, 1967; Meirovich and Baruh, 1983; Baruh and Choe, 1990), the system matrix is block diagonal with diagonal blocks representing second order oscillators. *The weak coupling control theory is a promising tool in the study of nonlinear mechanical systems and in general, systems with distributed parameters.* We hope that the methodology of this chapter can be extended to the general nonlinear systems (Khalil, 1992) as well.

The results of this chapter can be also extended to the problem of solving the bilinear-quadratic optimal control problems of weakly coupled systems by the successive approximations technique to be presented in Chapter 4. In

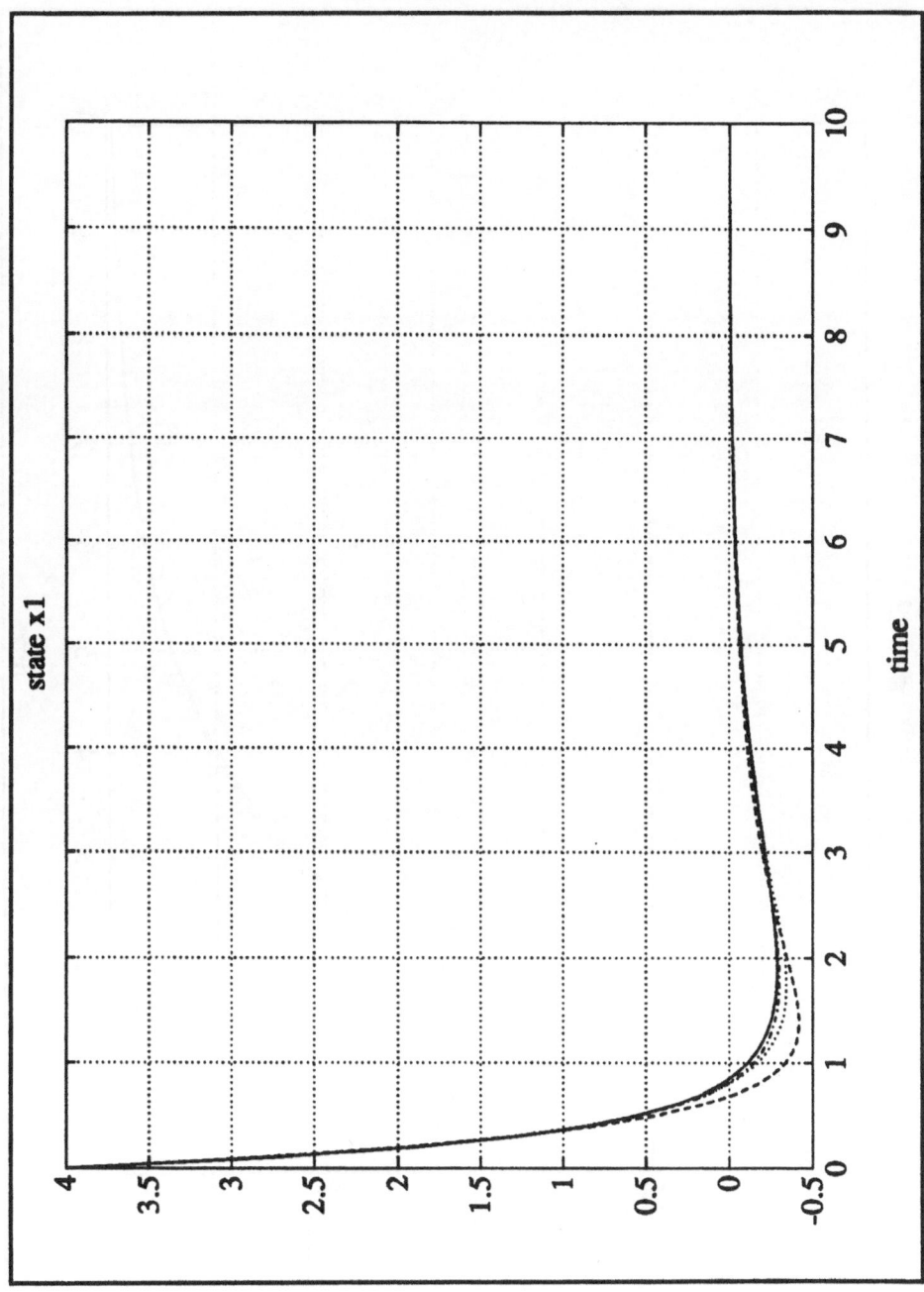

Figure 3.1: Optimal and approximate trajectories for x_1

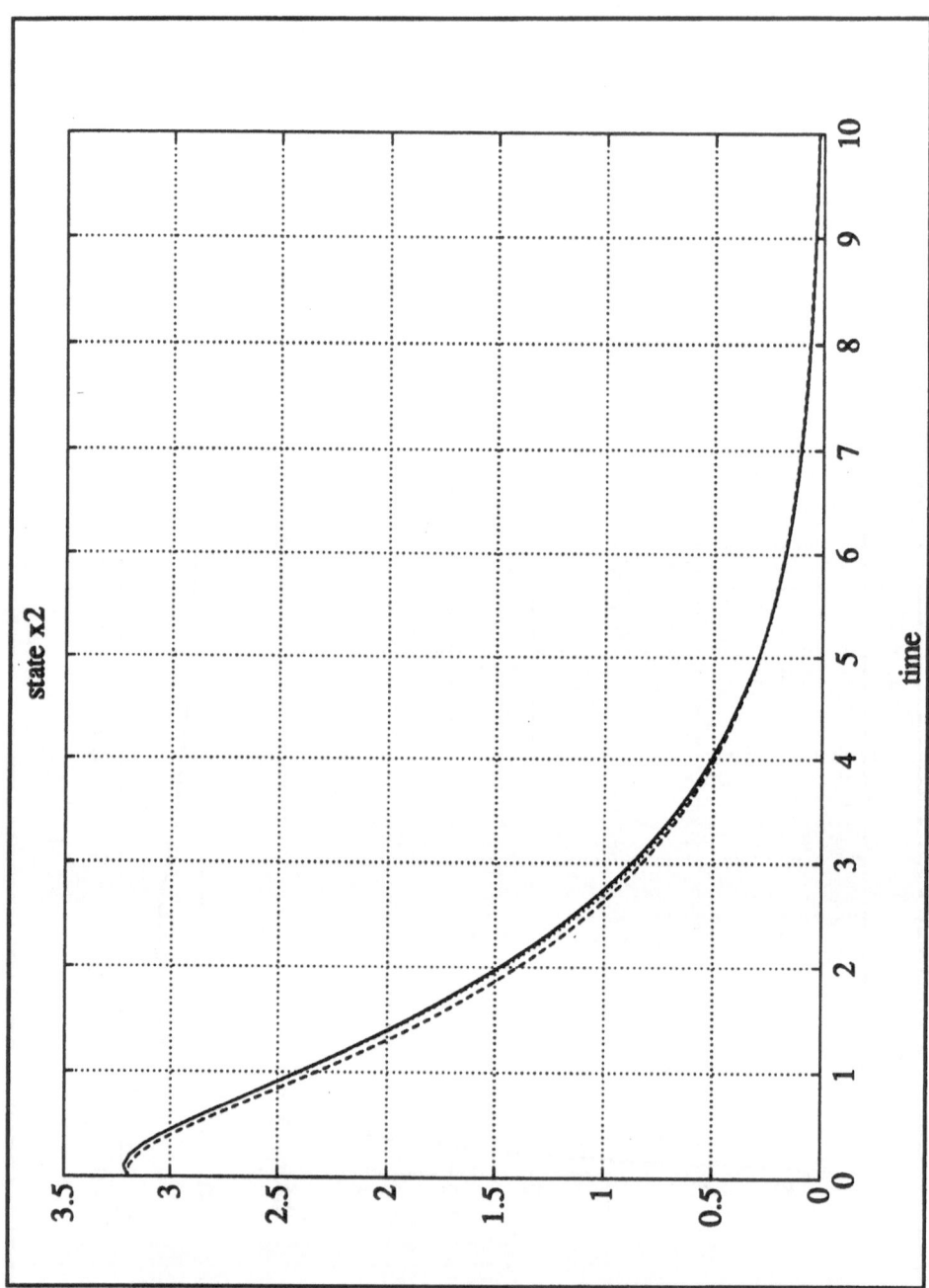

Figure 3.2: Optimal and approximate trajectories for x_2

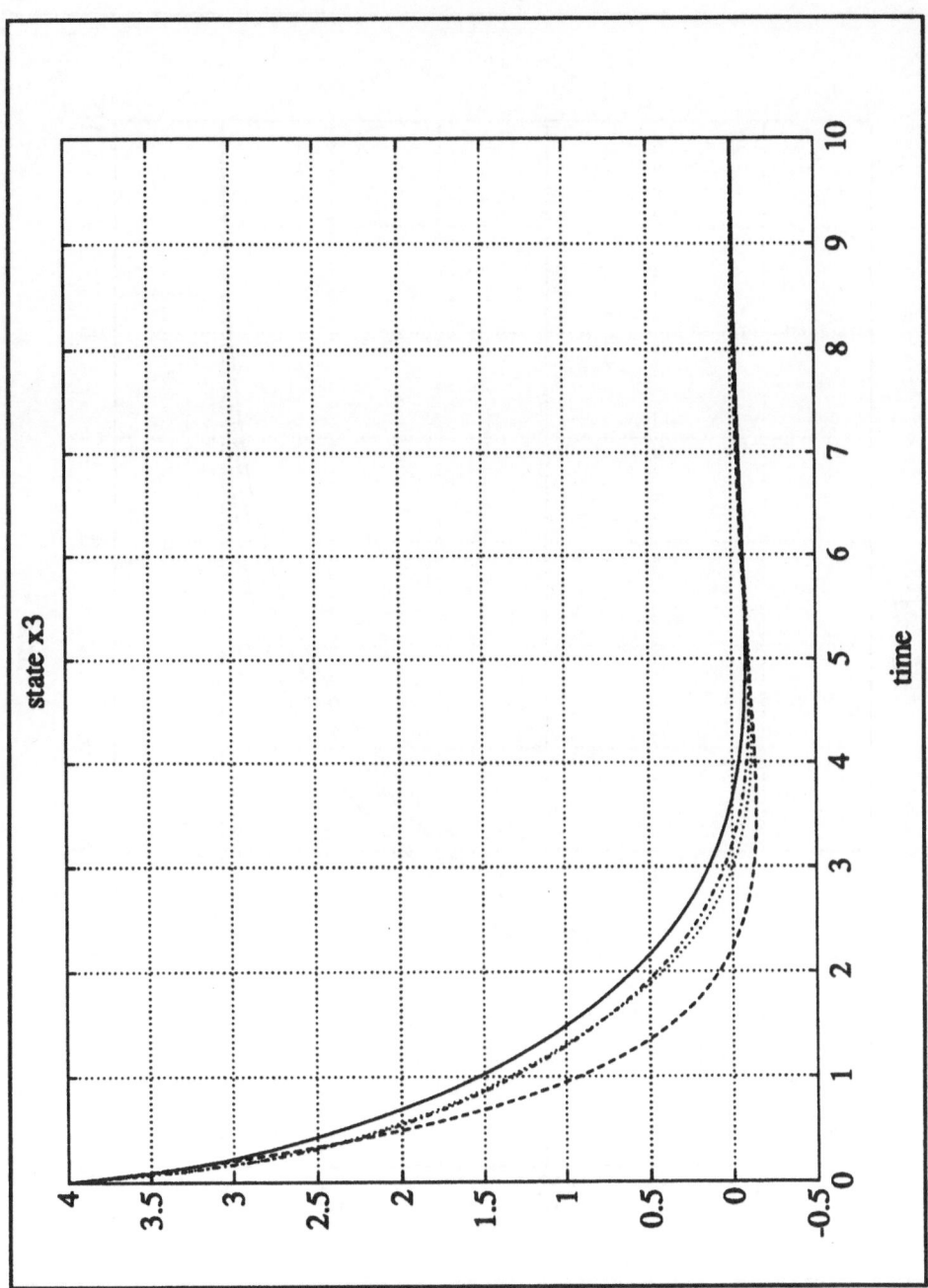

Figure 3.3: Optimal and approximate trajectories x_3

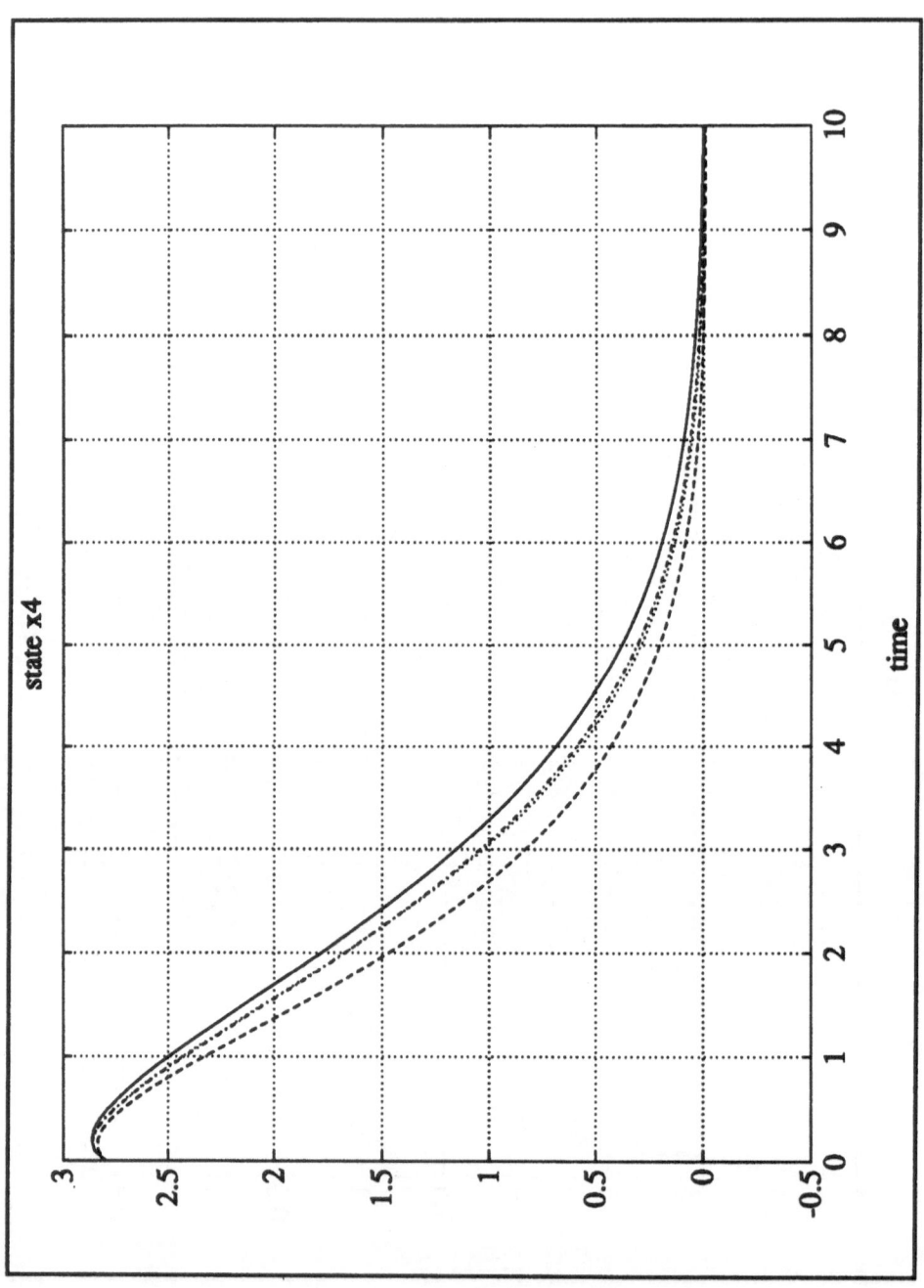

Figure 3.4: Optimal and approximate trajectories for x_4

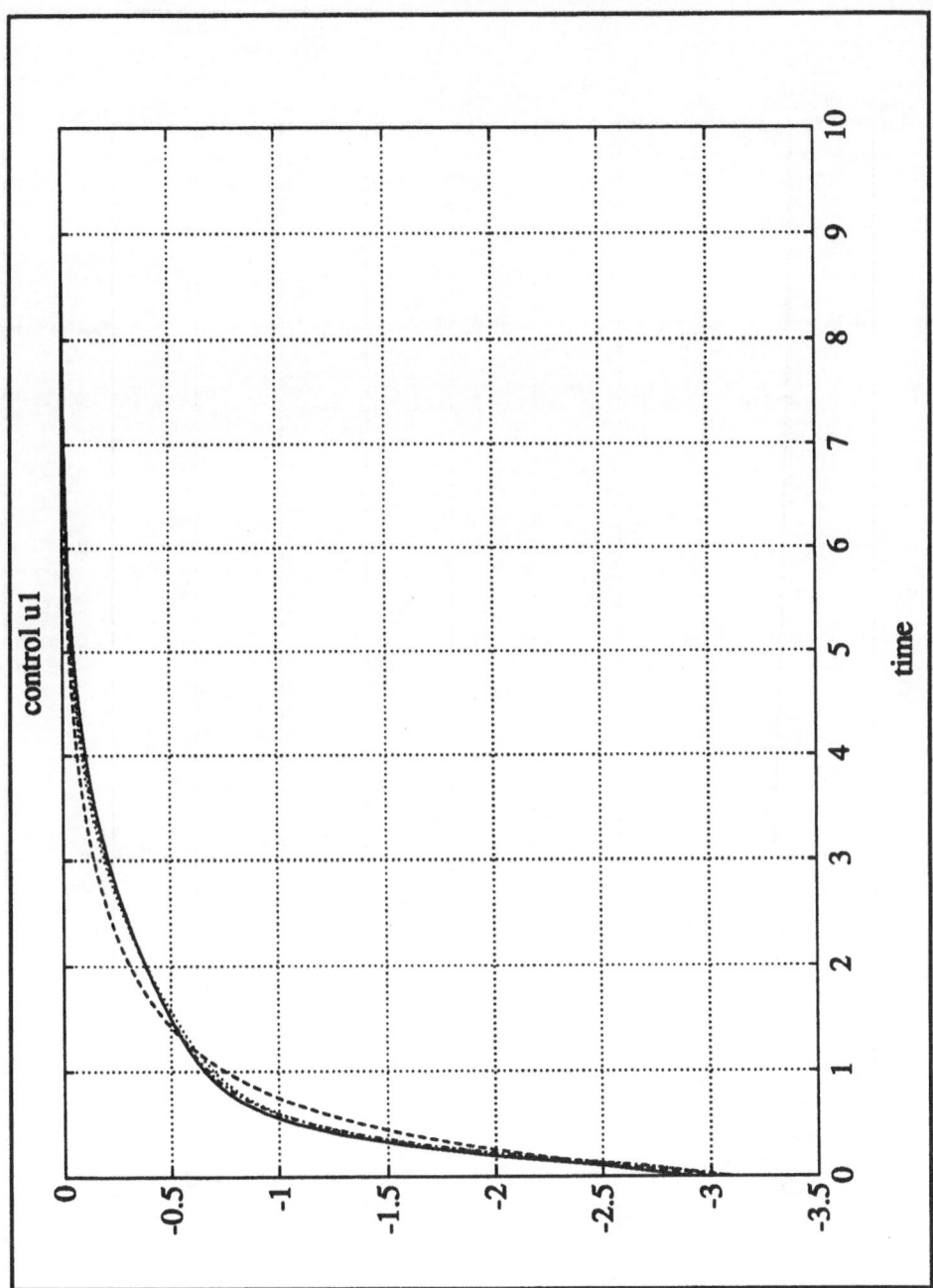

Figure 3.5: Optimal and approximate trajectories for control u_1

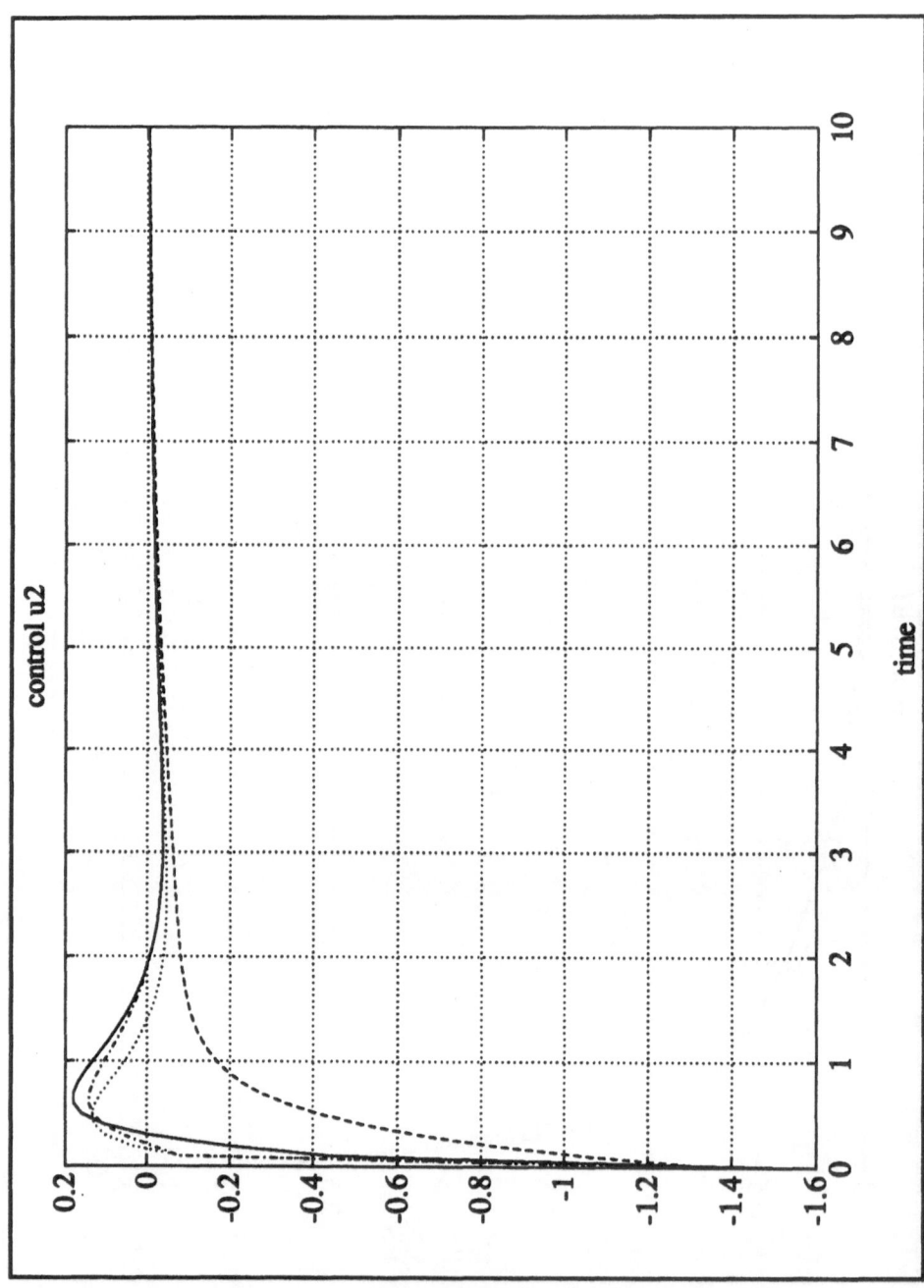

Figure 3.6: Optimal and approximate trajectories for control u_2

addition, as pointed out before, the study of the above problem in the discrete-time domain will facilitate some difficulties encountered in the continuous-time domain. This is left to the reader as an interesting research topic.

Appendix 3.1

Decoupling Transformations for Weakly Coupled Linear Systems

Consider a linear time invariant weakly coupled system

$$\dot{x}_1 = A_1 x_1 + \epsilon A_2 x_2 + B_1 u_1 + \epsilon B_2 u_2$$
$$\dot{x}_2 = \epsilon A_3 x_1 + A_4 x_2 + \epsilon B_3 u_1 + B_4 u_2 \tag{a.1}$$

Introducing the change of variables as (Gajic and Shen, 1989)

$$\eta_1 = x_1 - \epsilon L x_2 \tag{a.2}$$

the corresponding subsystem state equation in the new coordinates becomes

$$\dot{\eta}_1 = \left(A_1 - \epsilon^2 L A_3 \right) \eta_1 + \epsilon \left[A_1 L - L A_4 + A_2 - \epsilon^2 L A_3 L \right] x_2$$
$$+ \left(B_1 - \epsilon^2 L B_3 \right) u_1 + \epsilon (B_2 - L B_4) u_2 \tag{a.3}$$

By setting the coefficient multiplying x_2 in (a.3) to zero, we get an independent differential equation for η_1. This is possible since the obtained equation

$$A_1 L - L A_4 + A_2 - \epsilon^2 L A_3 L = 0 \tag{a.4}$$

has a solution, which in the case when A_1 and A_4 have no eigenvalues in common and for sufficiently small ϵ is unique (Gajic and Shen, 1989).

Introducing another change of variables as

$$\eta_2 = x_2 + \epsilon H \eta_1 \tag{a.5}$$

the other subsystem equation from (a.1) in the new coordinates is

$$\dot{\eta}_2 = \left(A_4 + \epsilon^2 A_3 L \right) \eta_2 + \epsilon \left[H \left(A_1 - \epsilon^2 L A_3 \right) - \left(A_4 + \epsilon^2 A_3 L \right) H + A_3 \right] \eta_1$$
$$+ \epsilon \left[B_3 + H \left(B_1 - \epsilon^2 L B_3 \right) \right] u_1 + \left[B_4 + \epsilon^2 H (B_2 - L B_4) \right] u_2 \tag{a.6}$$

This equation becomes independent of η_1 if we set the coefficient multiplying η_1 to zero, that is

$$H \left(A_1 - \epsilon^2 L A_3 \right) - \left(A_4 + \epsilon^2 A_3 L \right) H + A_3 = 0 \tag{3.48}$$

which is feasible and implies the unique solution for H when A_1 and A_4 have no eigenvalues in common and ϵ is sufficiently small. Thus, the transformation of (Gajic and Shen, 1989) is defined by (a.2) and (a.5), that is

$$\begin{bmatrix} \eta_1 \\ \eta_2 \end{bmatrix} = \begin{bmatrix} I & -\epsilon L \\ \epsilon H & I - \epsilon^2 H L \end{bmatrix} \begin{bmatrix} x_1 \\ x_2 \end{bmatrix} = T_2 \begin{bmatrix} x_1 \\ x_2 \end{bmatrix} \tag{a.8}$$

This nonsingular transformation has the inverse given by

$$T_2^{-1} = \begin{bmatrix} I - \epsilon^2 L H & \epsilon L \\ -\epsilon H & I \end{bmatrix} \tag{a.9}$$

Note that the same decomposition procedure is applicable in the case of time varying linear weakly coupled systems. In that case the algebraic equations (a.4) and (a.7) are replaced by the differential ones with no initial conditions imposed on L and H.

In (Qureshi, 1992) a new version of the Chang transformation is obtained by using the following change of variables

$$\begin{aligned} \eta_1 &= x_1 - \epsilon L_1 x_2 \\ \eta_2 &= -\epsilon H_1 x_1 + x_2 \end{aligned} \tag{a.10}$$

In addition of decomposing the system equations into independent subsystems, the transformation (a.10) also decomposes the transformation equations (a.4) and (a.7) so that they are independent of each other and can be solved in parallel.

In the new coordinates, the system (a.1) under transformation (a.10) becomes

$$\begin{aligned} \dot{\eta}_1 &= \left(A_1 - \epsilon^2 L_1 A_3\right)\eta_1 + B_{10}u_1 + \epsilon B_{20}u_2 \\ \dot{\eta}_2 &= \left(A_4 - \epsilon^2 H_1 A_2\right)\eta_2 + \epsilon B_{30}u_1 + B_{40}u_2 \end{aligned} \tag{a.11}$$

while the transformation equations are given by

$$\begin{aligned} L_1 A_4 - A_1 L_1 - A_2 + \epsilon^2 L_1 A_3 L_1 &= 0 \\ H_1 A_1 - A_4 H_1 - A_3 + \epsilon^2 H_1 A_2 H_1 &= 0 \end{aligned} \tag{a.12}$$

The time varying version of the new version of Chang transformation is derived in (Qureshi and Gajic, 1991; Qureshi, 1992). In the time varying case the only difference is that the algebraic transformation equations become differential ones with no initial conditions imposed.

Appendix 3.2

Hybrid Newton Fixed-Point Iterations Algorithm for the Algebraic Riccati Equation of Weakly Coupled Systems

In this appendix we present the hybrid Newton fixed-point iterations method for solving the algebraic Riccati equation of weakly coupled systems by following the work of (Rutkowski, 1995). Consider

$$A^T P + PA + Q - PSP = 0 \qquad (b.1)$$

with

$$A = \begin{bmatrix} A_1 & \epsilon A_2 \\ \epsilon A_3 & A_4 \end{bmatrix}, \quad B = \begin{bmatrix} B_1 & \epsilon B_2 \\ \epsilon B_3 & B_4 \end{bmatrix}$$

$$Q = \begin{bmatrix} Q_1 & \epsilon Q_2 \\ \epsilon Q_2^T & Q_3 \end{bmatrix}, \quad R = \begin{bmatrix} R_1 & 0 \\ 0 & R_2 \end{bmatrix}, \quad P = \begin{bmatrix} P_1 & \epsilon P_2 \\ \epsilon P_2^T & P_3 \end{bmatrix}$$

$$(b.2)$$

where ϵ is a small parameter indicating the separation of the state variables into two weakly coupled subsystems. A linear system with this partitioning displays the weakly coupled structure under the following assumption.

Assumption b.1 Assuming that the coefficient matrices are continuous functions of ϵ, then the separation (b.1)-(b.2) into subsystems is induced by the following property $det(A_1) = O(1)$ and $det(A_2) = O(1)$, (Chow and Kokotovic, 1983; Gajic and Shen, 1993).

$$\triangle$$

It is known that the Newton method is convenient for weakly coupled systems whenever the first-order approximation (obtained by setting $\epsilon = 0$ in the partitioned equation (b.1)) is close to the exact one (good initial guess). In that case the Newton method is superior over the fixed-point type reduced-order parallel algorithm for solving this Riccati equation (Gajic and Shen, 1993), owing to its quadratic speed of convergence. The numerical results will be obtained in terms of the reduced-order decoupled Lyapunov equations corresponding to the subsystems.

Substitution of (b.2) into (b.1) will produce the following equations

$$P_1 A_1 + A_1^T P_1 + Q_1 - P_1 S_1 P_1 + \epsilon^2 (P_2 A_3 + A_3^T P_2^T)$$
$$-\epsilon^2 [(P_1 S_{12} + P_2 Z^T) P_1 + (P_1 Z + P_2 (S_2 + \epsilon^2 S_{21})) P_2^T] = 0 \qquad \text{(b.3)}$$

$$P_3 A_4 + A_4^T P_3 + Q_3 - P_3 S_2 P_3 + \epsilon^2 (P_2^T A_2 + A_2^T P_2)$$
$$-\epsilon^2 [(P_3 S_{21} + P_2^T Z) P_3 + (P_3 Z + P_2^T (S_1 + \epsilon^2 S_{12}) P_2)] = 0 \qquad \text{(b.4)}$$

$$P_1 A_2 + P_2 A_4 + A_4^T P_2 + A_3^T P_3 + Q_2 - P_1 S_1 P_2$$
$$-P_1 S P_3 - \epsilon^2 [(P_1 S_{12} + P_2 Z^T) P_2 + P_2 S_{21} P_3] = 0 \qquad \text{(b.5)}$$

where

$$S_1 = B_1 R_1^{-1} B_1^T, \quad S_2 = B_4 R_2^{-1} B_4^T, \quad S_{12} = B_2 R_2^{-1} B_2^T$$
$$S_{21} = B_3 R_1^{-1} B_3^T, \quad Z = B_1 R_1^{-1} B_3^T + B_2 R_2^{-1} B_4^T \qquad \text{(b.6)}$$

The Newton method for solving nonlinear algebraic equations is based on the linearization procedure, namely, if we assume that $P_1^{(i)}, P_2^{(i)}, P_3^{(i)}$ are known points then the new iteration points are obtained as

$$P_j^{(i+1)} = P_j^{(i)} + \Delta P_j^{(i)}, \quad j = 1, 2, 3 \qquad \text{(b.7)}$$

where $\Delta P_j^{(i)}$ are small quantities. Substituting (b.7) in (b.3)-(b.6) and neglecting $O(\Delta^2)$ terms as very small ones (this is the linearization step of the Newton method) we obtain the following set of equations

$$P_1^{(i+1)} D_1^{(i)} + D_1^{(i)^T} P_1^{(i+1)} + \epsilon^2 \left(P_2^{(i+1)} D_3^{(i)} + D_3^{(i)^T} P_2^{(i+1)^T} \right) + Q_{11}^{(i)} = 0 \qquad \text{(b.8)}$$

$$P_3^{(i+1)} D_4^{(i)} + D_4^{(i)^T} P_3^{(i+1)} + \epsilon^2 \left(P_2^{(i+1)^T} D_2^{(i)} + D_2^{(i)^T} P_2^{(i+1)} \right) + Q_{33}^{(i)} = 0 \qquad \text{(b.9)}$$

$$P_1^{(i+1)} D_2^{(i)} + P_2^{(i+1)} D_4^{(i)} + D_3^{(i)^T} P_3^{(i+1)} + D_1^{(i)^T} P_2^{(i+1)} + Q_{22}^{(i)} = 0 \quad \text{(b.10)}$$

where

$$D_1^{(i)} = A_1 - S_1 P_1^{(i)} - \epsilon^2 \left(S_{12} P_1^{(i)} - Z P_2^{(i)^T} \right)$$

$$D_2^{(i)} = A_2 - Z P_3^{(i)} - (S_1 + \epsilon^2 S_{12}) P_2^{(i)}$$

(b.11)

$$D_3^{(i)} = A_3 - Z^T P_1^{(i)} - (S_2 + \epsilon^2 S_{21}) P_2^{(i)^T}$$

$$D_4^{(i)} = A_4 - S_2 P_3^{(i)} - \epsilon^2 \left(S_{21} P_3^{(i)} + Z^T P_2^{(i)} \right)$$

(b.11)

and

$$Q_{11}^{(i)} = P_1^{(i)} S_1 P_1^{(i)} + Q_1 +$$
$$\epsilon^2 \left[P_1^{(i)} S_{12} P_1^{(i)} + P_2^{(i)} Z^T P_1^{(i)} + P_1^{(i)} Z P_2^{(i)^T} + P_2^{(i)} (S_2 + \epsilon^2 S_{21}) P_2^{(i)^T} \right]$$

$$Q_{22}^{(i)} = Q_2 + P_1^{(i)} S_1 P_2^{(i)} + P_1^{(i)} Z P_3^{(i)} + P_2^{(i)} S_2 P_3^{(i)}$$
$$+ \epsilon^2 \left(P_1^{(i)} S_{12} P_2^{(i)} + P_2^{(i)} Z^T P_2^{(i)} + P_2^{(i)} S_{21} P_3^{(i)} \right)$$

$$Q_{33}^{(i)} = Q_3 + P_3^{(i)} S_2 P_3^{(i)} +$$
$$\epsilon^2 \left[P_3^{(i)} S_{21} P_3^{(i)} + P_2^{(i)^T} Z P_3^{(i)} + P_3^{(i)} Z^T P_2^{(i)} + P_3^{(i)^T} (S_1 + \epsilon^2 S_{12}) P_2^{(i)} \right]$$

(b.12)

Equations (b.8)-(b.10) can be decoupled by using the fixed-point iterations for their solution (Gajic and Shen, 1993). We start by setting $\epsilon = 0$ in (b.8)-(b.10). This produces

$$\mathbf{P}_1^{(i+1)} D_1^{(i)} + D_1^{(i)^T} \mathbf{P}_1^{(i+1)} + Q_{11}^{(i)} = 0$$

(b.13)

$$D_1^{(i)^T} \mathbf{P}_2^{(i+1)} + \mathbf{P}_2^{(i+1)} D_4^{(i)} + \mathbf{P}_1^{(i+1)} D_2^{(i)} + D_3^{(i)^T} \mathbf{P}_3^{(i+1)} + Q_{22}^{(i)} = 0 \quad \text{(b.14)}$$

$$\mathbf{P}_3^{(i+1)} D_4^{(i)} + D_4^{(i)^T} \mathbf{P}_3^{(i+1)} + Q_{33}^{(i)} = 0$$

(b.15)

Unique solutions to the first-order approximations (b.13)-(b.17) exist under the following assumption.

Assumption b.1 Matrices $D_1^{(i)}$ and $D_4^{(i)}$ are stable.

This standard assumption in the theory of weak coupling is always satisfied since $D_1^{(i)}$ and $D_4^{(i)}$ represent the closed-loop slow and fast subsystem matrices. The stability of the closed-loop matrices is the consequence of the stabilizability-detectability conditions imposed on subsystems.

In order to get solutions of (b.8)-(b.10) in terms of the solutions of (b.15)-(b.17) with any arbitrary order of accuracy, we define error terms as

$$P_j^{(i+1)} = \mathbf{P}_j^{(i+1)} + \epsilon^2 E_j, \quad j = 1, 2, 3 \tag{b.16}$$

Substituting (b.16) in (b.8)-(b.10) and using the fixed-point parallel algorithm from (Gajic and Shen, 1993) for solving the corresponding Lyapunov equations of the error terms, we get three completely decoupled linear equations

$$E_1^{(k+1)} D_1^{(i)} + D_1^{(i)^T} E_1^{(k+1)} = -D_3^{(i)^T} \mathbf{P}_2^{(i+1)^T} - \mathbf{P}_2^{(i+1)} D_3^{(i)} \tag{b.17}$$

$$E_3^{(k+1)} D_4^{(i)} + D_4^{(i)^T} E_3^{(k+1)} = -D_2^{(i)^T} \mathbf{P}_2^{(i+1)} - \mathbf{P}_2^{(i+1)} D_2^{(i)} \tag{b.18}$$

$$E_2^{(k+1)} D_4^{(i)} + D_1^{(i)^T} E_2^{(k+1)} = -D_3^{(i)^T} E_3^{(k+1)} - E_1^{(k+1)} D_2^{(i)} \tag{b.19}$$

with the starting points chosen as $E_1^{(0)} = 0$, $E_2^{(0)} = 0$, $E_3^{(0)} = 0$.

The fixed-point algorithm defined in (b.17)-(b.19) converges to the exact solution of the error terms with the rate of convergence of $O(\epsilon^2)$, that is (Gajic and Shen, 1993)

$$\left\| E_j - E_j^{(k)} \right\| = O\left(\epsilon^{2k}\right), \quad k = 1, 2, ... \tag{b.20}$$

Using results from (b.20) in (b.16) implies that after k iterations we get

$$P_j^{(i+1)} = \mathbf{P}_j^{(i+1)} + \epsilon^2 E_j^{(k)} + O\left(\epsilon^{2(k+1)}\right) \tag{b.21}$$

After obtaining $P_j^{(i+1)}$ with the desired accuracy we go to the next iteration step with respect to i. Note that the i iterations are the Newton method iterations so that one has to perform only 4–5 iterations with respect to i (the Newton method either converges in 4–5 iterations or does not converge at all).

Numerical examples that demonstrate the efficiency of the hybrid Newton fixed-point iterations algorithm for solving the weakly coupled algebraic Riccati equation can be found in (Rutkowski, 1995).

Chapter 4

The Successive Approximation Procedure for Optimal Control of Bilinear Systems

In this chapter we present the solution to the optimum regulation problem of a bilinear system with a quadratic performance criterion. The solution is obtained in terms of a sequence of the algebraic Lyapunov equations. The results are based on the method of successive approximations. Both the steady state and finite time optimization problems are considered. The proofs of convergence of the presented schemes are given and the design procedures are illustrated by examples. The presented results are mostly based on the work reported in (Aganovic, 1993; Aganovic and Gajic, 1995).

4.1 Introduction

From the practical point of view there is a need for the application oriented controller design technique for bilinear systems. However, for a bilinear system and a standard quadratic cost functional, with the exception of the simplest cases, it is not possible to express the optimal control in the explicit feedback form. Most of the obtained results rely on quadratic cost functionals modified by inclusions of additional nonnegative state-dependent penalizing functions. An overview of the available results can be found in (Mohler,

1970, 1973, 1991; Bruni et al., 1974; Benallou et al., 1988). Some of the available results are presented in Section 2.2. The obtained optimal controls have problems with global stabilization of the closed-loop system and with physical meaning of the modified cost functionals.

A new line of thought in the optimization of bilinear systems has been the development of an approximative procedure (Cebuhar and Constanza, 1984). The algorithm obtained is characterized by a sequence of linear-quadratic problems converging to the overall optimal solution. Since the process of the actual computation of the approximate control is still numerically complicated, it is the purpose of this chapter to present a new iterative scheme, based on the method of successive approximations, which produces linear "feedback" control law which is simpler to compute than the one obtained in (Cebuhar and Constanza, 1984).

It is well established that direct attempts to solve the Hamilton-Jacobi-Bellman equation, in order to obtain the optimal feedback controls, are hopeless in all but a few special cases. Bellman foresaw these difficulties in his work on dynamic programming (Bellman, 1957), and in addition to his main constructive method of solution laid strong emphasis on the use of successive approximations for the study of optimal processes.

This chapter is organized as follows. In Section 4.2, the approximate method of (Cebuhar and Constanza, 1988) is presented in detail. In Section 4.3, we use the method of successive approximations in order to derive a new iterative scheme which is characterized by the problem of solving a sequence of time-varying algebraic Lyapunov equations. The proof of convergence of the new method is given. The numerical example that illustrates the efficiency of the proposed iterative scheme is presented in Section 4.4.

In the remaining sections we study the finite time optimization problem of bilinear-quadratic control systems (Hofer and Tibken, 1988). It is shown that the successive approximation procedure simplifies computations of the optimal solution of the bilinear-quadratic optimal control problem. On the contrary of the results of Hofer and Tibken where the optimal solution has been obtained in terms of a sequence of the differential Riccati equations in the presented method only solutions of a sequence of the differential Lyapunov equations are required. A chemical reactor example is used to demonstrate the efficiency of the new method.

4.2 Steady State Bilinear-Quadratic Control Problem

Consider the optimal control problem of a bilinear multi-input multi-output system defined by

$$\dot{x} = Ax + Bu + \{xN\}u,$$

$$x(0) = x_0, \quad \{xN\} = \sum_{j=1}^{n} x_j N_j \tag{4.1}$$

where $x \in \Re^n$ are system state variables, $u \in \Re^m$ are control inputs, A, B, and N_j are constant matrices of appropriate dimensions with $N_j \in \Re^{n \times m}$.

The quadratic cost functional to be minimized is given by

$$J = \frac{1}{2} \int_{t_0}^{\infty} \left(x^T Q x + u^T R u\right) dt \tag{4.2}$$

where Q is a positive semidefinite symmetric $n \times n$ matrix and R a positive definite symmetric $m \times m$ matrix. The above steady state optimization problem has been considered in (Cebuhar and Constanza, 1984) under the following assumption.

Assumption 4.1 The pair (A, B) is completely controllable and x stays in the controllability domain defined by

$$X_c \triangleq \{x \in \Re^n | (A, B + xN) \text{ controllable}\}$$

$$\triangle$$

Since the results of (Cebuhar and Constanza, 1984) are based on (Jacobson, 1980) it is important to state the main assumption from (Jacobson, 1980) as well. It is formulated as.

Assumption 4.2 The differential equation (4.1) has a solution defined on $[0, +\infty)$ for each admissible input function and $x(t) \to 0$ as $t \to \infty$.

$$\triangle$$

Dynamic programming approach applied to (4.1)-(4.2) results in the steady state Hamilton-Jacobi-Bellman equation of the form, (Cebuhar and Constanza, 1984)

$$\frac{1}{2}x^T Q x + (J_x)^T A x - \frac{1}{2}(J_x)^T (B + \{xN\}) R^{-1} (B + \{xN\})^T (J_x) = 0 \tag{4.3}$$

Since it is known (Jacobson, 1980) that for the infinite-time case $J = J(x)$, then the solution of the Hamilton-Jacobi-Bellman equation can be sought in the form $p(x) = J_x^* = P(x)x$, where the matrix-valued function $P(x)$ is symmetric (Jacobson, 1980). The equation (4.3) is now reduced to

$$Q + P(x)A + A^T P(x) - P(x)(B + \{xN\})R^{-1}(B + \{xN\})^T P(x) = 0 \tag{4.4}$$

The required optimal control is in the form

$$u^{opt} = -R^{-1}(B + \{xN\})^T P(x)x \tag{4.5}$$

Unfortunately, there is no analytical solution to equation (4.4). Therefore, there is a need for finding an approximate method for solving the optimal control problem of bilinear systems.

It was shown in (Bruni et al., 1971) that the solutions of a sequence of linear differential equations

$$\dot{x}_0 = Ax_0 + Bu, \quad x_0(t_0) = x^0 \tag{4.6}$$

$$\dot{x}_i = Ax_i + \{x_{i-1}N\}u + Bu, \quad x_i(t_0) = x^0, \quad i = 1, 2, ... \tag{4.7}$$

converge uniformly under the same control input u to the solution x of (4.1). By using this result and Assumptions 4.1 and 4.2, the optimization problem of the bilinear system (4.1) subject to (4.2) is replaced by a sequence of linear-quadratic optimization problems given by, (Cebuhar and Constanza, 1984)

$$\dot{x}_i = Ax_i + B_i(t)u, \quad B_i(t) \triangleq B + \{x_{i-1}^*(t)N\} \tag{4.8}$$

with the algebraic Riccati equations

$$Q + P_i(t)A + A^T P_i(t) - P_i(t)B_i(t)R^{-1}B_i(t)^T P_i(t) = 0, \quad i = 1, 2, ... \tag{4.9}$$

and the corresponding optimal controls

$$u_i^* = -R^{-1}B_i(t)^T P_i(t)x_i^* \tag{4.10}$$

An asterisk indicates the optimal quantities. A schematic diagram of this algorithm is presented in Figure 4.1.

The trajectories x_i^* are solutions of (4.8), with u_i^* in place of u. It has been proved in (Cebuhar and Constanza, 1984) that if x^* and u^* represent the solutions of the optimization problem (4.1) and (4.2), then the sequence $\{x_i^*\}$ converges uniformly to the optimal state trajectory x^* and the sequence $\{u_i^*\}$ converges uniformly to the optimal control u^*.

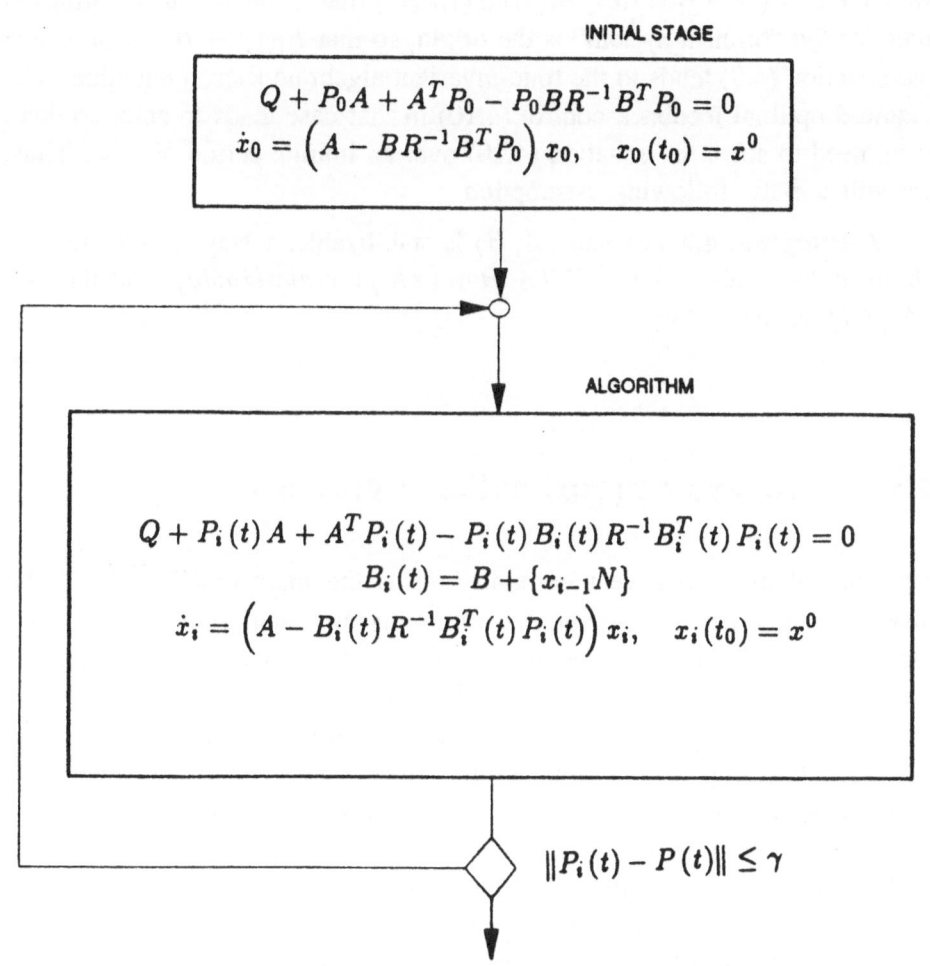

INITIAL STAGE

$$Q + P_0 A + A^T P_0 - P_0 B R^{-1} B^T P_0 = 0$$
$$\dot{x}_0 = \left(A - B R^{-1} B^T P_0 \right) x_0, \quad x_0(t_0) = x^0$$

ALGORITHM

$$Q + P_i(t) A + A^T P_i(t) - P_i(t) B_i(t) R^{-1} B_i^T(t) P_i(t) = 0$$
$$B_i(t) = B + \{x_{i-1}N\}$$
$$\dot{x}_i = \left(A - B_i(t) R^{-1} B_i^T(t) P_i(t) \right) x_i, \quad x_i(t_0) = x^0$$

$$\|P_i(t) - P(t)\| \leq \gamma$$

Figure 4.1: Convergent sequence of linear-quadratic optimal control problems

In this section, we will relax the controllability assumption of (Cebuhar and Constanza, 1984) into the stabilizability assumption. Also, since the matrix Q in (4.9) does not change per iteration it is convenient to assume that the pair $(A, \sqrt{Q})$ is detectable. This will lead to the existence of the unique stabilizing solution $P_i(t)$, in order words, the matrix $A - B_i(t)R^{-1}B_i(t)P_i(t)$ will be asymptotically stable for every frozen $t \in [0, \infty)$. Due to stability of the closed-loop system matrix, at steady state we have $0 = \left(A - B_i(t)R^{-1}B_i(t)P_i(t)\right)x_e(t)$ that is, the unique equilibrium point of the "bilinear system" is the origin, so that $B_i(t) \to B = const$, and the equation (4.9) tends to the time-invariant algebraic Riccati equation. The required optimal feedback control (4.10) in that case tends to zero, so there is no need to solve the equation (4.10) over an infinite period of time. Thus, we will use the following assumption.

Assumption 4.3 The pair (A, B) is stabilizable, x stays in the stabilizability domain $X_s = \{x \in \Re^n | (A, B + \{xN\})\ stabilizable\}$, and the pair $(A, \sqrt{Q})$ is detectable.

$\triangle$

4.3. Successive Approximation Approach

The method of successive approximations is the main tool in solving the functional equation of dynamic programming (Bellman, 1954, 1957, 1961; Larson, 1967). It has been used in several control theory papers, for example, (Vaisbord, 1963; Kleinman, 1968; Levine and Vilis, 1973; Mageriou, 1977; Mil'shtein 1964). This method can be used as a very powerful decomposition technique that simplifies computations. The monotonicity of successive approximations can be easily established as shown in (Bellman, 1961, page 171). However, proving the convergence is a much more complex task. In the work of (Mil'shtein, 1964), an approximate convergent method for synthesis of the optimal control system is investigated. The approach is based on a combination of the ideas of Lyapunov's second method and Bellman's method of successive approximations. Convergent suboptimal control sequences were also obtained in (Vaisbord, 1963; Kleinman, 1968; Mageriou, 1977; Leake and Liu, 1967).

The first step in developing the new optimization algorithm is based on the application of the method of successive approximations to the approximative procedure presented in (Cebuhar and Constanza, 1984). As a consequence of this we will be faced with the problem of solving the algebraic Lyapunov equations instead the algebraic Riccati equations as required in (Cebuhar and Constanza, 1984). Since the complete derivation of the successive approximation scheme for the linear-quadratic control problem can not be found in the literature in the explicit form, we have produced corresponding derivations in Appendix 4.1.

The successive approximation technique results in the following computational steps.

Step 1: For the known continuously differentiable and stabilizable control $u_{ki}(x_i)$ find the value of the performance criterion

$$J_{ki}(x_i, u_{ki}(x_i)) = \frac{1}{2} \int_{t_0}^{\infty} \left(x_i^T Q x_i + u_{ki}^T R u_{ki} \right) dt \qquad (4.11)$$

along the trajectories of the dynamical system (4.8) driven by $u_{ki}(x_i)$.

$\triangle$

Step 2: For the known value of the performance criterion $J_{ki}(x_i)$ find a new approximation of the optimal control $u_{(k+1)i}(x_i)$ by minimization of the partially frozen "Hamiltonian"

$$\min_{u \in U} \left[\left(\frac{\partial J_{ki}}{\partial x_i}, (A x_i + B_i(t)u) \right) + \left(x_i^T Q x_i + u^T R u \right) \right] \qquad (4.12)$$

Note that $\frac{\partial J_{ki}}{\partial x_i}$ is frozen, that is, known from the previous iteration.

$\triangle$

If we assume that the known stabilizable control $u_{ki}(x_i)$ can be expressed in the linear feedback form as (see Appendix 4.1)

$$u_{ki} = -R^{-1} B_i^T(t) S_{(k-1)i}(t) x_i(t) \qquad (4.13)$$

so that from the minimization of (4.12) we get

$$u_{(k+1)i} = -R^{-1} B_i^T(t) \frac{\partial J_{ki}(x_i)}{\partial x_i} = -R^{-1} B_i^T(t) S_{ki}(t) x_i(t) \qquad (4.14)$$

then the actual computation of $\frac{\partial J_{ki}}{\partial x_i}$ can be obtained in the form (see Appendix 4.1)

$$\frac{\partial J_{ki}}{\partial x_i} = S_{ki}(t)x_i(t) \tag{4.15}$$

where $S_{ki}(t)$ is a symmetric matrix. After some mathematical derivations, by using the procedure outlined in Appendix 4.1, Steps 1 and 2 of the successive approximations produce the following algorithm.

Algorithm 4.1:

Solve the Lyapunov equations

$$S_{ki}(t)A_{ki}(t) + A_{ki}^T(t)S_{ki}(t) + Q_{ki}(t) = 0 \tag{4.16}$$

where

$$A_{ki}(t) = A - B_i(t)R^{-1}B_i^T(t)S_{(k-1)i}(t)$$

$$Q_{ki}(t) = Q + S_{(k-1)i}(t)B_i(t)R^{-1}B_i^T(t)S_{(k-1)i}(t) \tag{4.17}$$

$$B_i(t) = B + \{x_{i-1}N\}$$

The approximations are initialized at each iteration step by $S_{0i}(t) = S_{k(i-1)}(t)$, a symmetric and positive definite matrix, which stabilizes the system

$$\dot{x}_i = \left(A - B_i(t)R^{-1}B_i(t)^T S_{0i}(t)\right)x_i, \quad x(t_o) = x^0 \tag{4.18}$$

In the first step of the sequence of linear systems, the approximation procedure is initialized by a matrix S_{00} that stabilizes the system

$$\dot{x}_0 = \left(A - BR^{-1}B^T S_{00}\right)x_0, \quad x_0(t_o) = x^0 \tag{4.19}$$

Then, the iterative procedure

$$\begin{aligned} S_{k0}A_{k0} + A_{k0}^T S_{k0} + Q_{k0} &= 0 \\ A_{k0} &= A - BR^{-1}B^T S_{(k-1)0} \\ Q_{k0} &= Q + S_{(k-1)0}BR^{-1}B^T S_{(k-1)0} \end{aligned} \tag{4.20}$$

is performed. One way to find an initial stabilizing matrix S_{00} would be to use the results of (Kleinman, 1970). An alternative way for the above initial stage procedure is to use the positive semidefinite stabilizing solution of the following algebraic Riccati equation

$$Q + S_{00}A + A^T S_{00} - S_{00}BR^{-1}B^T S_{00} = 0 \qquad (4.21)$$

$$\triangle$$

So far, as a result of the application of the idea of successive approximations, we have managed to replace the computation of the time-varying algebraic Riccati equation by a sequence of the time-varying algebraic Lyapunov equations (see Figure 4.2). This scheme reminds us of the famous Kleinman algorithm (Kleinman, 1968) for the approximate solution of the time-invariant Riccati algebraic equation. As a matter of fact the Kleinman algorithm can be derived from the theory of successive approximations.

The solutions of the Lyapunov equations converge to the solution of (4.4)

$$S_{ki}(t) \to P_i(t), \qquad k \to \infty \qquad (4.22)$$

and since we proceed to the next step of the approximation scheme (4.3)-(4.4) only after obtaining $S_{ki}(t)$ close enough to $P_i(t)$

$$\|S_{ki}(t) - P_i(t)\| \le \gamma, \qquad \gamma \quad \text{small enough} \qquad (4.23)$$

we except that convergence of the procedure (4.3)-(4.4) is preserved. We also expect that the optimization error $O(\gamma)$ which propagate through a sequence of linear systems leads to

$$x_i^{opt}(t) = x_i(t) + O(\gamma) \qquad (4.24)$$

Since the above algorithm represents just an intermediate result in the process of the development of a new procedure, the proof of its convergence is omitted.

Having in mind the above results of the application of the successive approximations, simulations results showing that both iteration loops $S_{ki}(t) \to P_i(t)$ and $P_i(t) \to P(t)$ are characterized by the rapid convergence, and the analytical convergence proof from (Cebuhar and Constanza, 1984) the following question is raised: Do we need to perform all of the iterations from (4.16)-(4.17) in order to obtain the desired accuracy (4.23) at

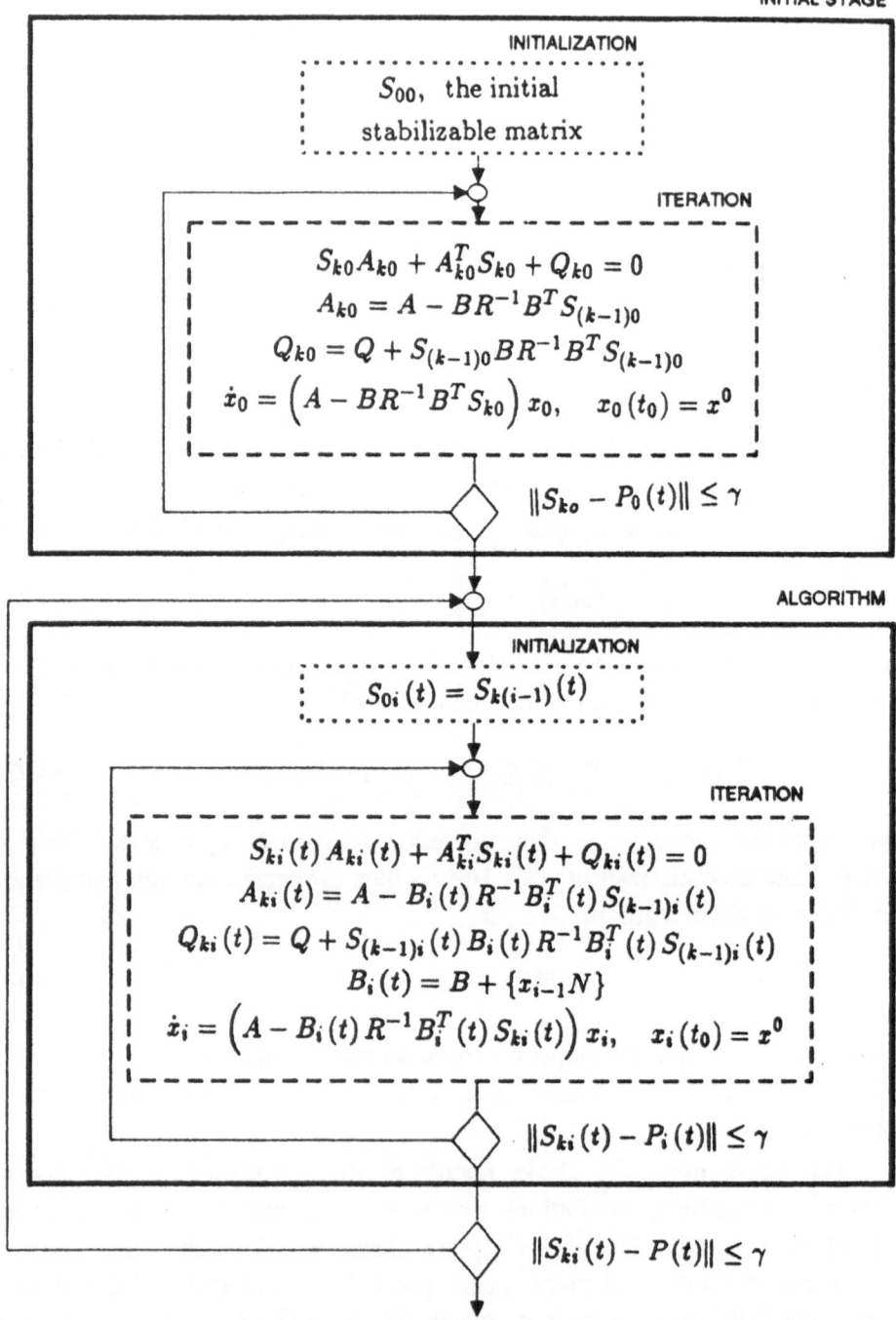

Figure 4.2: Successive approximations and sequence of linear problems

each linear sequence level? The answer is no. The modified and simplified version of Algorithm 4.1 is our new algorithm. It is given in the following form.

Algorithm 4.2:

Solve the Lyapunov equations

$$S_i(t)A_i(t) + A_i^T(t)S_i(t) + Q_i(t) = 0 \qquad (4.25)$$

where

$$A_i(t) = A - B_i(t)R^{-1}B_i^T(t)S_{i-1}(t)$$

$$Q_i(t) = Q + S_{i-1}(t)B_i(t)R^{-1}B_i^T(t)S_{i-1}(t) \qquad (4.26)$$

$$B_i(t) = B + \{x_{i-1}N\}$$

and

$$\dot{x}_i(t) = \left(A - B_i(t)R^{-1}B_i^T(t)S_i(t)\right)x_i(t), \qquad x_i(t_0) = x^0 \qquad (4.27)$$

The sequence (4.25)-(4.27) is initialized by S_0, the real symmetric positive definite matrix which is the solution of the following algebraic Riccati equation

$$Q + S_0 A + A^T S_0 - S_0 B R^{-1} B^T S_0 = 0 \qquad (4.28)$$

This solution stabilizes the system

$$\dot{x}_0 = \left(A - B R^{-1} B^T S_0\right)x_0, \qquad x(t_0) = x^0 \qquad (4.29)$$

$$\Delta$$

This new iteration scheme replaces the solution of the time-varying algebraic Riccati equation by the solution of the time-varying algebraic Lyapunov equation at each iteration step, Figure 4.3.

A summary of numerical methods for efficiently solving the Lyapunov equations can be found in (Gajic and Qureshi, 1995).

The simulation results show that both iteration loops $S_i(t) \rightarrow P_i(t)$ and $P_i(t) \rightarrow P(t)$ are compatible and that the inner iteration loop for obtaining desired accuracy (4.23) is redundant. Namely, after only a few iteration steps

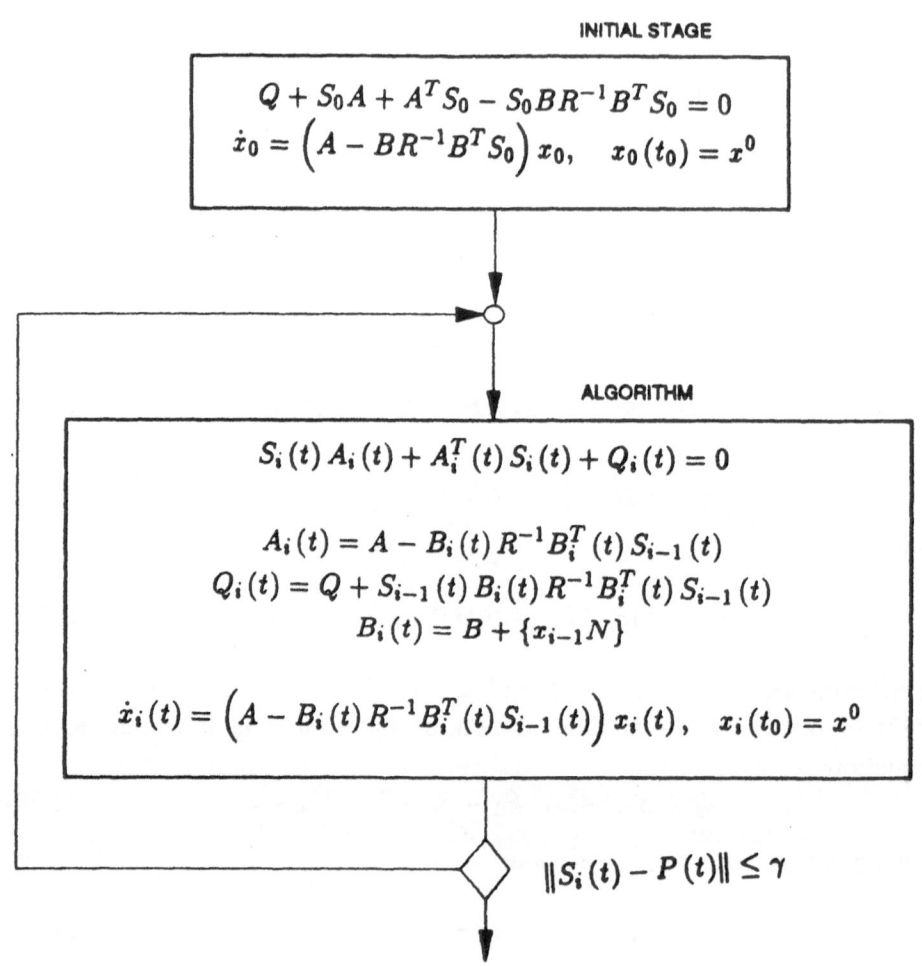

Figure 4.3: Optimization as sequence of the Lyapunov equations

$S_i(t)$, the solution of the time-varying algebraic Lyapunov equation of the iteration scheme (4.25)-(4.27), gets equally close to the optimal solution $P(t)$ of (4.4) as so does $P_i(t)$ — the solution of the time-varying algebraic Riccati equation of the iterative scheme (4.8)-(4.9).

4.3.1 Proof of Convergence

In this section, we give the convergence proof for Algorithm 4.2. The proof is along the line of the corresponding one from (Cebuhar and Constanza, 1984) taking into account the specific features of the newly proposed algorithm. The trajectories x_i^* are the solutions of (4.27) and the trajectories x^* are the solutions of

$$\dot{x} = \left[A - (B + \{xN\})R^{-1}(B + \{xN\})^T S\right]x, \qquad x(t_0) = x^0 \qquad (4.30)$$

where $S_i(t)$ is obtained from (4.25) and $S(t)$ satisfies

$$S(t)\widehat{A}(x) + \widehat{A}^T(x)S(t) + \widehat{Q}(x) = 0 \qquad (4.31)$$

The above equation is the equivalent of (4.4) with $S(t) = P(x(t))$, $\widehat{A}(x)$, and $\widehat{Q}(x)$ defined as

$$\widehat{A}(x) = A - B(x)R^{-1}B^T(x)S(t), \qquad B(x) = B + \{xN\}$$

$$(4.32)$$

$$\widehat{Q}(x) = Q + S(t)B(x)R^{-1}B^T(x)S(t)$$

The following difference between $x(t)$ and $x_i(t)$ is easily obtained from (4.27) and (4.30)

$$
\begin{aligned}
\frac{d}{dt}(x - x_i) = {}& \widehat{A}(x)(x - x_i) - \{[B(x)R^{-1}B^T(x) \\
& - B_i(x_{i-1})R^{-1}B_i^T(x_{i-1})]S(t) \\
& + B_i(x_{i-1})R^{-1}B_i^T(x_{i-1})[S(t) - S_i(t)]\}x_i
\end{aligned}
\qquad (4.33)
$$

Using variation of constants method and the notion of the system transition matrix we have

$$\frac{d}{dt}\phi(t, t_0) = \widehat{A}(x)\phi(t, t_0), \qquad \phi(t_0, t_0) = I$$

$$\frac{d}{dt}\phi_i(t, t_0) = \left(A - B_i(x_{i-1})R^{-1}B_i^T(x_{i-1})S_i(t)\right)\phi_i(t, t_0) \qquad (4.34)$$

$$\phi_i(t_0, t_0) = I$$

which implies the following expression for the difference of considered state trajectories $x^*(t) - x_i^*(t)$

$$z_i(t) \triangleq x^*(t) - x_i^*(t) = \int_{t_0}^{t} \phi(t,s)\{[B(x)R^{-1}B(x) - B_i(x_{i-1})R^{-1}$$

$$\times B_i^T(x_{i-1})]S(s) + B_i(x_{i-1})R^{-1}B_i^T(x_{i-1})[S(s) - S_i(s)]\}\phi_i(t,s)x^0 ds$$
(4.35)

Taking the norm of both sides and using the norm properties leads to

$$\|z_i(t)\| \le \int_{t_0}^{t} \{\beta_1\|B(x)R^{-1}B^T(x) - B_i(x_{i-1})R^{-1}B_i^T(x_{i-1})\|$$
(4.36)

$$+\beta_2\|S(s) - S_i(s)\|\}ds$$

where β_1 and β_2 are obtained by straightforward calculations from (4.35)

$$\beta_1 = \|x^0\|\|\phi(t,s)\|\|S(s)\|\|\phi_i(t,s)\|$$
(4.37)

$$\beta_2 = \|x^0\|\|\phi(t,s)\|\|B_i(x_{i-1})R^{-1}B_i^T(x_{i-1})\|\|\phi_i(t,s)\|$$

It is important to notice that by Assumptions 4.2 and 4.3 all norms defined in (4.37) are bounded.

We can also choose constants M_1 and M_2 such that, for any bounded $x \in X_c$

$$\|B(x)R^{-1}B^T(x) - B_i(x_{i-1})R^{-1}B_i^T(x_{i-1})\| \le M_1\|x^* - x_{i-1}^*\|$$
(4.38)

$$\|S(s) - S_i(s)\| \le M_2\|x^* - x_{i-1}^*\|$$

The last result is due to the analyticity of the solution of the algebraic Riccati and Lyapunov equations, (Cebuhar and Constanza, 1984; Ran and Rodman, 1988).

Therefore

$$\|z_i(t)\| \le \int_{t_0}^{t} (\beta_1 M_1 + \beta_2 M_2)\|z_{i-1}(s)\|ds$$
(4.39)

or

$$\|z_i(t)\| \le M \int_{t_0}^{t} \|z_{i-1}(s)\| ds \qquad (4.40)$$

Since M can be chosen independently of i, by recursion, we have

$$\|z_i(t)\| \le M^i \int_{t_0}^{t} \int_{t_0}^{s_1} \cdots \int_{t_0}^{s_{i-1}} \|z_0(s_i)\| ds_i ... ds_1 \qquad (4.41)$$

and, since z_0 is bounded on $[0, T]$ (choose an upper bound as N), it follows that

$$\|z_i(t)\| \le \left[(MT)^i/i!\right] N, \quad \forall t \in [0, T] \qquad (4.42)$$

As i tends to infinity, the ratio $(MT)^i/i!$ tends to zero; therefore, $z_i(t)$ converges to zero, uniformly over $t \in [0, T]$, $0 < T < \infty$, which is the identical result as one obtained in (Cebuhar and Constanza, 1984). Since by Assumptions 4.2 and 4.3 both $x_i^*(t) \to 0$ and $x^*(t) \to 0$ as $t \to \infty$, then $\|z_i(t)\| \to 0$ for all $t \in [0, +\infty)$.

4.4. Examples

In this section we consider two example, the first one already available in the literature and the second one a real physical bilinear model of a paper making machine.

4.4.1 A General Bilinear System

For the sake of comparison, the example done in (Cebuhar and Constanza, 1984) is used to illustrate the efficiency of the new iterative scheme. The system and cost functional are given as follows

$$\begin{bmatrix} \dot{x}_1 \\ \dot{x}_2 \end{bmatrix} = \begin{bmatrix} -1 & 4 \\ -4 & -1 \end{bmatrix} \begin{bmatrix} x_1 \\ x_2 \end{bmatrix} + x_1 \begin{bmatrix} 1 & 0 \\ 1 & 1 \end{bmatrix} \begin{bmatrix} u_1 \\ u_2 \end{bmatrix}$$

$$+ x_2 \begin{bmatrix} 1 & 0 \\ 1 & 1 \end{bmatrix} \begin{bmatrix} u_1 \\ u_2 \end{bmatrix} + \begin{bmatrix} 3 & 1 \\ 0 & 2 \end{bmatrix} \begin{bmatrix} u_1 \\ u_2 \end{bmatrix}$$

$$J = \frac{1}{2} \int\limits_0^\infty \left(2x_1^2 + x_2^2 + u_1^2 + 2u_1 u_2 + 4u_2^2 \right) dt$$

The simulation results for the proposed Algorithm 4.2 are given in Figures 4.4–4.5 and Table 4.1. In Figures 4.4–4.5 we present the state trajectories. It can be seen that the new procedure requires only 2 iterations in order to get the optimal trajectories. Since in our case we are solving the Lyapunov equations instead of the Riccati equations, as was required in (Cebuhar and Constanza, 1984), the computational time is obviously reduced for the algorithm proposed.

state [x1]

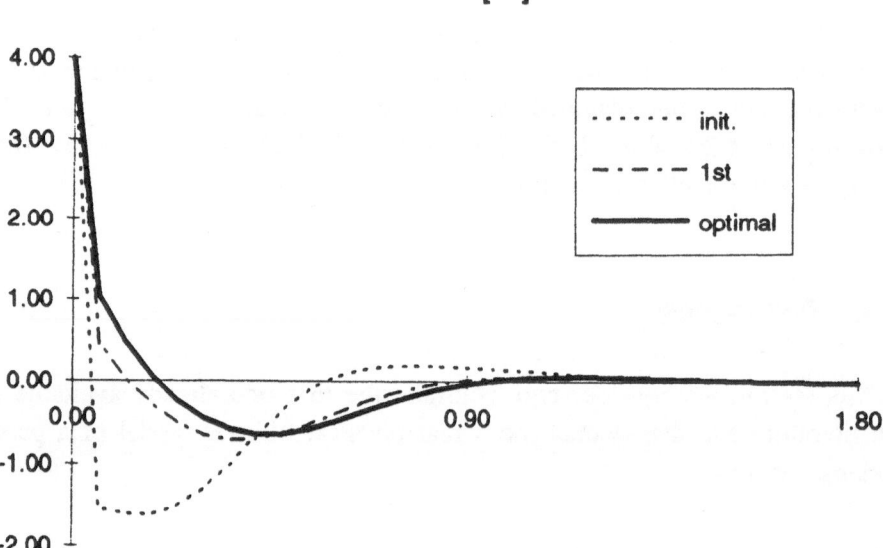

time [s]

Figure 4.4: Profiles of state x_1 in iteration procedure (4.25)-(4.28)

In Table 4.1 we present the performance criterion per iteration. In the same table for the sake of comparison we present the performance criterion per iteration for the method of (Cebuhar and Constanza, 1984). It can be seen from Table 4.1 that the speed of convergence of these two algorithms is comparable, but again, Algorithm 4.2 requires far less computations, and thus, generates the required optimal solution faster.

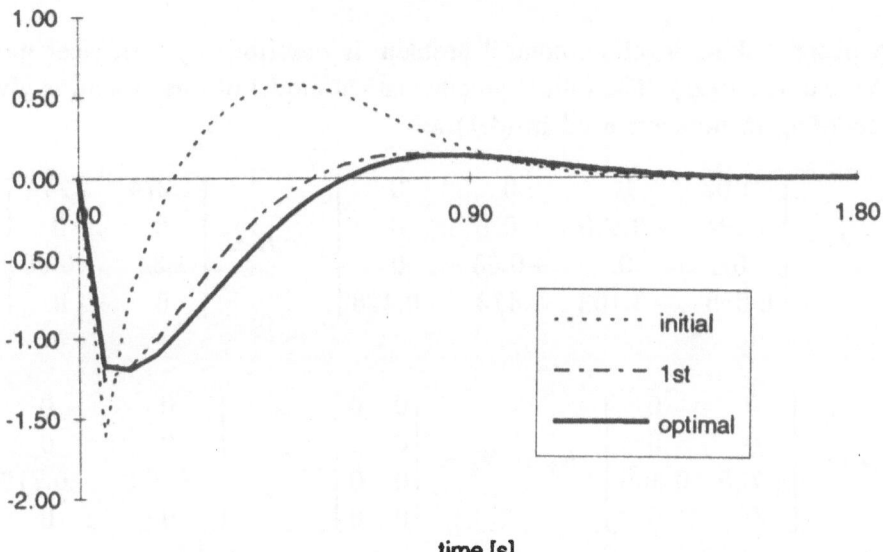

Figure 4.5: Profiles of state x_2 in iteration procedure (4.25)-(4.28)

	Algorithm 4.2	*Algorithm of (Chebuhar and Constanza, 1984)*
0	5.9018	5.9018
1	2.8167	2.4732
2	2.4941	2.4722
3	2.4719	2.4717
4 = optimal	2.4717	2.4717

Table 4.1: Performance criteria per iteration for two algorithms

4.4.2 Case Study: A Paper Making machine

A paper making machine control problem is described by a bilinear model (Ying et al., 1992). The bilinear mathematical model of this system is given, according to notation used in (4.1) as

$$A = \begin{bmatrix} -1.93 & 0 & 0 & 0 \\ 0.394 & -0.426 & 0 & 0 \\ 0 & 0 & -0.63 & 0 \\ 0.095 & -0.103 & 0.413 & -0.426 \end{bmatrix}, \quad B = \begin{bmatrix} 1.274 & 1.274 \\ 0 & 0 \\ 1.34 & -0.65 \\ 0 & 0 \end{bmatrix}$$

$$N_1 = \begin{bmatrix} 0 & 0 \\ 0 & 0 \\ 0.755 & 0.366 \\ 0 & 0 \end{bmatrix}, N_2 = N_4 = \begin{bmatrix} 0 & 0 \\ 0 & 0 \\ 0 & 0 \\ 0 & 0 \end{bmatrix}, N_3 = \begin{bmatrix} 0 & 0 \\ 0 & 0 \\ -0.718 & -0.718 \\ 0 & 0 \end{bmatrix}$$

Weighting matrices Q and R chosen as

$$Q = \begin{bmatrix} 1 & 0 & 0.13 & 0 \\ 0 & 1 & 0 & 0.09 \\ 0.13 & 0 & 0.1 & 0 \\ 0 & 0.09 & 0 & 0.2 \end{bmatrix}, \quad R = \begin{bmatrix} 1 & 0 \\ 0 & 1 \end{bmatrix}$$

The simulation results for the proposed Algorithm 4.2 are given in Figures 4.6–4.7 for the optimal and approximate trajectories and in Table 4.2 for the optimal and approximate performance criteria. We present only results for state trajectories x_3 and x_4 since the results for x_1 and x_2 are even better.

It can be seen from Figures 4.6–4.7 and Table 4.2 that on this real physical example the proposed algorithm also performs very well, that is, it produces quite quickly very good approximations for the optimal state trajectories.

Note that for this particular real physical bilinear system the obtained results are extra ordinary good and very close to the results obtained by using the linear model owning to the fact that the matrices $N_i, i = 1, 2, 3, 4$, are very sparse so that the impact of the multiplicative control is not fully utilized.

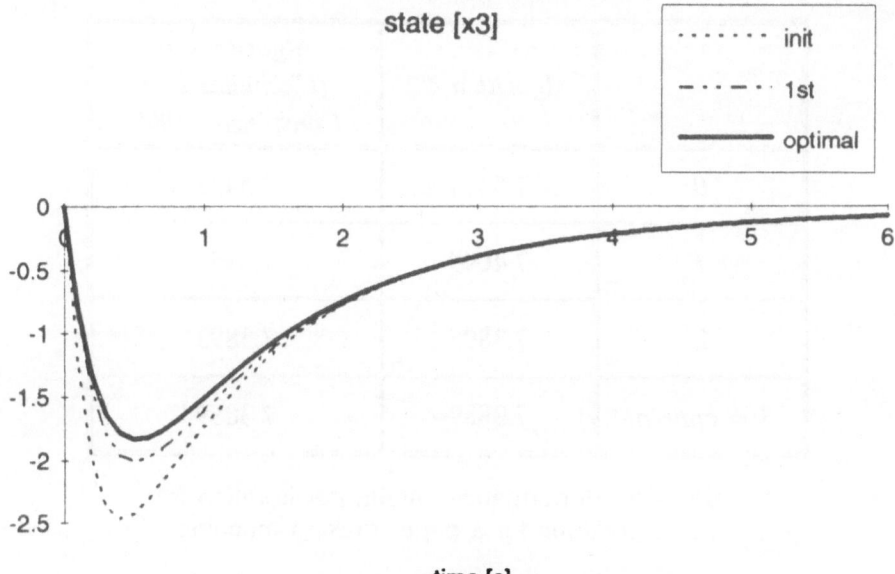

Figure 4.6: Profiles of state x_3 in iteration
procedure (4.25)-(4.28) for a paper making machine

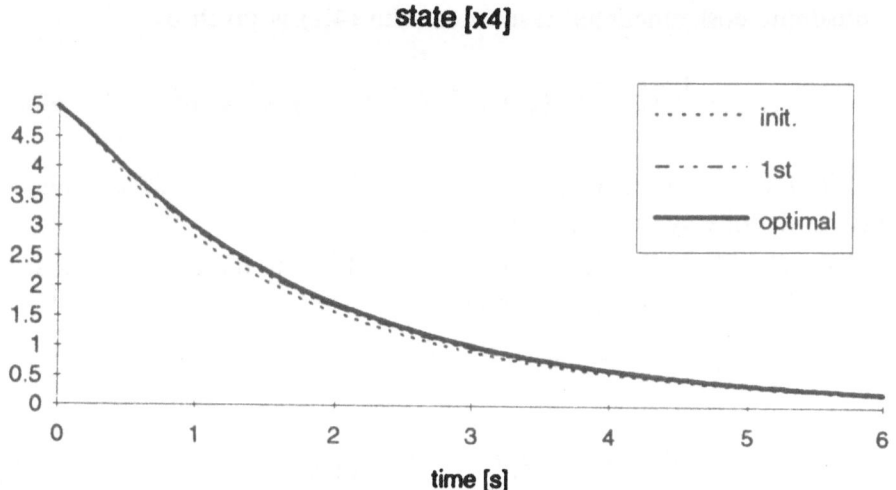

Figure 4.7: Profiles of state x_4 in iteration
procedure (4.25)-(4.28) for a paper making machine

	Algorithm 4.2	Algorithm of (Chebuhar and Constanza, 1984)
0	7.5431	7.5431
1	7.4049	7.3957
2	7.3895	7.3893
3 = optimal	7.3889	7.3889

Table 4.2: Performance criteria per iteration for two algorithms for a paper making machine

4.5 Finite-Time Optimal Control of Bilinear Systems

Consider the finite-time optimal control problem of the bilinear system (4.1). The quadratic cost functional associated with (4.1) is given by

$$J = \frac{1}{2}x(t_f)^T F x(t_f) + \frac{1}{2}\int_{t_0}^{t_f} (x^T Q x + u^T R u)\, dt \qquad (4.43)$$

where Q and F are positive semidefinite symmetric $n \times n$ matrices, and R a positive definite symmetric $m \times m$ matrix.

The application of the minimum principle leads to the following nonlinear two-point boundary value system

$$\dot{x}_i = [Ax]_i - \left[(B + \{xN\})R^{-1}(B + \{xN\})^T p\right]_i, \quad x_i(t_0) = x_i^0$$

$$\dot{p}_i = -[Qx]_i - [A^T p]_i + \frac{1}{2}p^T\{N_i R^{-1}(B + \{xN\})^T \qquad (4.44)$$

$$+ (B + \{xN\})R^{-1}N_i^T\}p, \qquad p_i(t_f) = [Fx(t_f)]_i$$

where $[\,.\,]_i$, $i = 1, ..., n$, stands for the i-th component of the corresponding vector. Unfortunately, there is no analytical solution to this nonlinear two-point boundary value problem. Therefore, there is a need for finding an approximate method for solving the optimal control problem of bilinear systems.

The work of (Hofer and Tibken, 1988) introduces the iterative scheme that stays in close proximity to the Riccati approach of the linear-quadratic optimization. Namely, the state-costate system (4.44) is rewritten in the same form as in the linear case

$$\dot{x} = \widetilde{A}x - \widetilde{B}R^{-1}\widetilde{B}^T p, \qquad x(t_0) = x^0$$

$$\dot{p} = -\widetilde{Q}x - \widetilde{A}^T p, \qquad p\left(t^f\right) = Fx\left(t^f\right) \tag{4.45}$$

where newly introduced time-varying matrices $\widetilde{A}$, $\widetilde{Q}$, $\widetilde{B}R^{-1}\widetilde{B}^T$ are represented by the following expressions

$$\widetilde{A}_{ij} := A_{ij} - \frac{1}{2}[(N_j R^{-1}B^T + BR^{-1}N_j^T)p]_{.i}, \qquad i,j = 1, ..., n$$

$$\widetilde{Q}_{ij} := Q_{ij} - \frac{1}{2}p^T(N_i R^{-1}N_j^T + N_j R^{-1}N_i^T)p, \qquad i,j = 1, ..., n \tag{4.46}$$

$$\widetilde{B}R^{-1}\widetilde{B}^T := (B + \{xN\})R^{-1}(B + \{xN\})^T$$
$$- \frac{1}{2}\Big(\{xN\}R^{-1}B^T + BR^{-1}\{xN\}^T\Big)$$

Using (4.46) and denoting the iteration index by $k = 0, 1, ...$ and taking into account that

$$\widetilde{A}^{(k)} = \widetilde{A}\Big(p^{(k)}(t)\Big), \qquad \widetilde{Q}^{(k)} = \widetilde{Q}\Big(p^{(k)}(t)\Big)$$

$$\widetilde{B}^{(k)}R^{-1}\widetilde{B}^{(k)^T} = \widetilde{B}\Big(x^{(k)}(t)\Big)R^{-1}\widetilde{B}^T\Big(x^{(k)}(t)\Big) \tag{4.47}$$

the iterative solution of the state-costate equation (4.45) can be obtained as (Hofer and Tibken, 1988)

$$\dot{x}^{(k+1)} = \widetilde{A}^{(k)} x^{(k+1)} - \widetilde{B}^{(k)} R^{-1} \widetilde{B}^{(k)^T} p^{(k+1)}, \quad x^{(k+1)}(t_0) = x^0$$

$$\dot{p}^{(k+1)} = -\widetilde{Q}^{(k)} x^{(k+1)} - \widetilde{A}^{(k)^T} p^{(k+1)}, \tag{4.48}$$

$$p^{(k+1)}(t_f) = F x^{(k+1)}(t_f)$$

The iteration steps in (4.48) are carried out by using the Riccati formalism, that is

$$\dot{K}^{(k+1)} = -\widetilde{Q}^{(k)} - K^{(k+1)} \widetilde{A}^{(k)} - \widetilde{A}^{(k)^T} K^{(k+1)}$$
$$\quad + K^{(k+1)} \widetilde{B}^{(k)} R^{-1} \widetilde{B}^{(k)^T} K^{(k+1)}, \quad K^{(k+1)}(t_f) = F$$

$$\dot{x}^{(k+1)} = \left[\widetilde{A}^{(k)} - \widetilde{B}^{(k)} R^{-1} \widetilde{B}^{(k)^T} K^{(k+1)} \right] x^{(k+1)}, \quad x^{(k+1)}(t_0) = x^0 \tag{4.49}$$

Then, for each iteration step, the feedback controller is obtained as, (Hofer and Tibken, 1988)

$$u^{(k+1)}(t) = -R^{-1} \widetilde{B}^{(k)^T} K^{(k+1)}(t) x^{(k+1)}(t) \tag{4.50}$$

where the gain matrix $K^{(k+1)}(t)$ has to be calculated iteratively from the Riccati matrix differential equation (4.49). It was proven in (Hofer and Tibken, 1988) that convergence of this iterative scheme is guaranteed under the following assumption.

Assumption 4.4 The control penalty matrix R is large enough.

$$\Delta$$

The first step in developing a new optimization algorithm is based on the application of the method of successive approximations to the approximative procedure presented in (Hofer and Tibken, 1988). The idea is to use *only one* iteration of the successive approximation iterations at each step of the optimization procedure of (Hofer and Tibken, 1988). As a consequence of this we will have to solve only one differential Lyapunov equation at each iteration step. About the theory and numerical methods of differential

Lyapunov equation the interested reader can consult (Gajic and Qureshi, 1995). Since the derivation of the successive approximation scheme for the finite-time linear-quadratic optimal control problem can not be found in the literature in the explicit form we have produced complete derivations in Appendix 4.2. The convergence proof of the new iterative scheme will be given in the next section. Here, we present the algorithm only.

Equations defined in (4.48) correspond to the following linear-quadratic finite-time time-varying control problem

$$\dot{x}^{(k+1)} = \tilde{A}^{(k)} x^{(k+1)} + \tilde{B}^{(k)} u^{(k+1)}, \qquad x^{(k+1)}(t_0) = x^0 \tag{4.51}$$

$$J^{(k+1)} = \frac{1}{2} x^{(k+1)^T}(t_f) F x^{(k+1)}(t_f)$$
$$+ \frac{1}{2} \int_{t_0}^{t_f} \left(x^{(k+1)^T} \tilde{Q}^{(k)} x^{(k+1)} + u^{(k+1)^T} R u^{(k+1)} \right) dt \tag{4.52}$$

The one-step application of the successive approximation technique to (4.51)-(4.52), according to derivations outlined in Appendix 4.2, results in the following algorithm.

Algorithm 4.3:

$$\dot{x}^{(k+1)} = \left[\tilde{A}^{(k)} - \tilde{B}^{(k)} R^{-1} \tilde{B}^{(k)^T} P^{(k)} \right] x^{(k+1)} = A^{(k)} x^{(k+1)}$$
$$x^{(k+1)}(t_0) = x^0 \tag{4.53}$$

$$\dot{P}^{(k+1)} + P^{(k+1)} A^{(k)} + A^{(k)^T} P^{(k+1)} + Q^{(k)} = 0, \qquad P^{(k+1)}(t_f) = F \tag{4.54}$$

where

$$A^{(k)} = \tilde{A}^{(k)} - \tilde{B}^{(k)} R^{-1} \tilde{B}^{(k)^T} P^{(k)} = \tilde{A}^{(k)} - \tilde{S}^{(k)} P^{(k)}$$

$$Q^{(k)} = \tilde{Q}^{(k)} + P^{(k)} \tilde{B}^{(k)} R^{-1} \tilde{B}^{(k)^T} P^{(k)} = \tilde{Q}^{(k)} + P^{(k)} \tilde{S}^{(k)} P^{(k)} \tag{4.55}$$

For the first iteration step $k = 0$, the matrices $\widetilde{A}^{(0)}$, $\widetilde{B}^{(0)} R^{-1} \widetilde{B}^{(0)^T}$, and $\widetilde{Q}^{(0)}$ are calculated by using the solution of

$$\dot{x}^{(0)} = \left(A - B R^{-1} B^T P^{(0)} \right) x^{(0)}, \quad x^{(0)}(t_0) = x^0$$

$$\dot{P}^{(0)} + P^{(0)} A + A^T P^{(0)} + Q = 0, \quad P^{(0)}(t_f) = F \tag{4.56}$$

which corresponds to the linear part $\dot{x} = Ax + Bu$ of the bilinear system (4.1).

$\triangle$

Thus, the one-step application of the successive approximations requires the iterative solution of the time-varying differential Lyapunov equations, on the contrary to (Hofer and Tibken, 1988) where the solution of the differential time-varying Riccati equations is required at each iteration.

The approximate control law is stabilizable and given by

$$u^{(k)}\left(x^{(k+1)} \right) = -R^{-1} \left\{ \widetilde{B}^{(k)} \right\}^T P^{(k+1)} x^{(k+1)}$$

It is important to notice that in the proposed scheme we have to solve only one Lyapunov differential equation at each iteration. Namely, after obtaining the solution of the first Lyapunov differential equation we update all coefficients and go to the next iteration with respect to k. In that respect, the proposed method is a combination of the successive approximations and the scheme developed by Hofer and Tibken. In the next step, we have to prove the convergence of the proposed method. The proof is along the lines of (Hofer and Tibken, 1988) taking into account the specific features of the successive approximations.

4.5.1 Proof of Convergence

In the first part of this proof the expressions for the differences of $x^{(k+1)}(t) - x^{(k)}(t)$ and $P^{(k+1)}(t) - P^{(k)}(t)$ will be derived.

From (4.53) and (4.55) it can be obtained

$$\frac{d}{dt}\left(x^{(k+1)} - x^{(k)} \right) = A^{(k)}\left(x^{(k+1)} - x^{(k)} \right) + \left(A^{(k)} - A^{(k-1)} \right) x^{(k)} \tag{4.57}$$

By using the variation of constants and the definition the system transition matrix

$$\frac{d}{dt}\phi^{(k+1)} = A^{(k)}\phi^{(k+1)}, \quad \phi^{(k+1)}(0) = I, \quad x^{(k+1)}(t) = \phi^{(k+1)}(t)x^0$$

$$(4.58)$$

the expression for the difference $x^{(k+1)} - x^{(k)}$ can be written as

$$x^{(k+1)}(t) - x^{(k)}(t) =$$

$$\phi^{(k+1)}(t)\int_{t_0}^{t} \phi^{(k+1)^{-1}}(s)\Big(A^{(k)}(s) - A^{(k-1)}(s)\Big)\phi^{(k)}(s)x^0 ds$$

$$(4.59)$$

Similarly, from (4.54)-(4.55), it can be obtained

$$\frac{d}{dt}\Big(P^{(k+1)} - P^{(k)}\Big) + \Big(P^{(k+1)} - P^{(k)}\Big)A^{(k)} + A^{(k)^T}\Big(P^{(k+1)} - P^{(k)}\Big)$$

$$+ Q^{(k)} - Q^{(k-1)} + P^{(k)}\Big(A^{(k)} - A^{(k-1)}\Big) + \Big(A^{(k)} - A^{(k-1)}\Big)^T P^{(k)} = 0$$

$$(4.60)$$

so that the corresponding difference is

$$P^{(k+1)}(t) - P^{(k)}(t) = \Big(\phi^{(k+1)^{-T}}(t)\Big)$$

$$\times \{\int_{t_0}^{t_f} \phi^{(k+1)^T}(s)[\Big(Q^{(k)}(s) - Q^{(k-1)}(s)\Big)$$

$$+ P^{(k)}(s)\Big(A^{(k)}(s) - A^{(k-1)}(s)\Big)$$

$$(4.61)$$

$$+ \Big(A^{(k)}(s) - A^{(k-1)}(s)\Big)P^{(k+1)}(s)]\phi^{(k+1)}(s)ds\}\phi^{(k+1)^{-1}}(t)$$

Taking the norm of both sides of (4.59) and (4.61), we get

$$\left\|x^{(k+1)}(t) - x^{(k)}(t)\right\| \leq \int_{t_0}^{t_f} \alpha_1 \left\|A^{(k)}(s) - A^{(k-1)}(s)\right\| ds$$

$$\left\|P^{(k+1)}(t) - P^{(k)}(t)\right\| \leq \int_{t_0}^{t_f} \{\beta_1 \left\|A^{(k)}(s) - A^{(k-1)}(s)\right\|$$

$$(4.62)$$

$$+ \beta_2 \left\|Q^{(k)}(s) - Q^{(k-1)}(s)\right\|\}ds$$

where $\alpha_1, \beta_1, \beta_2$ are obtained by straightforward calculation from (4.59) and (4.61). For example α_1 is given by

$$\alpha_1 = \left\| x^0 \right\| \left\| \phi^{(k+1)}(t) \phi^{(k+1)^{-1}}(s) \right\| \left\| \phi^{(k)}(s) \right\| \tag{4.63}$$

In the next step we obtained the estimates of the norms for $\left\| A^{(k)}(t) - A^{(k-1)}(t) \right\|$ and $\left\| Q^{(k)}(t) - Q^{(k-1)}(t) \right\|$ in terms of the norms

$\left\| x^{(k)}(t) - x^{(k-1)}(t) \right\|$ and $\left\| P^{(k)}(t) - P^{(k-1)}(t) \right\|$. From (4.55) the following norm estimates can be obtained

$$\left\| A^{(k)} - A^{(k-1)} \right\| \le \left\| \widetilde{A}^{(k)} - \widetilde{A}^{(k-1)} \right\| + \left\| \widetilde{S}^{(k)} - \widetilde{S}^{(k-1)} \right\| \left\| P^{(k)} \right\|$$
$$+ \left\| \widetilde{S}^{(k-1)} \right\| \left\| P^{(k)} - P^{(k-1)} \right\| \tag{4.64}$$

$$\left\| Q^{(k)} - Q^{(k-1)} \right\| \le \left\| \widetilde{Q}^{(k)} - \widetilde{Q}^{(k-1)} \right\| + \left\| P^{(k)} - P^{(k-1)} \right\| \left\| \widetilde{S}^{(k)} P^{(k)} \right\| +$$

$$\left\| P^{(k-1)} \right\| \left\| \widetilde{S}^{(k)} - \widetilde{S}^{(k-1)} \right\| \left\| P^{(k)} \right\| + \left\| P^{(k-1)} \widetilde{S}^{(k-1)} \right\| \left\| P^{(k)} - P^{(k-1)} \right\|$$
$$\tag{4.65}$$

The norms of $\left\| \widetilde{A}^{(k)} - \widetilde{A}^{(k-1)} \right\|$, $\left\| \widetilde{S}^{(k)} - \widetilde{S}^{(k-1)} \right\|$, and $\left\| \widetilde{Q}^{(k)} - \widetilde{Q}^{(k-1)} \right\|$

can be estimated in terms of the original problem matrices (4.1), (4.43)-(4.47) so that the results of (Hofer and Tibken, 1988) can be used, that is

$$\left\| \widetilde{A}^{(k)} - \widetilde{A}^{(k-1)} \right\| \le \left[\sum_{j=1}^{n} \left\| \frac{1}{2} (N_j R^{-1} B^T + B R^{-1} N_j^T) \right\|^2 \right]^{\frac{1}{2}} \tag{4.66}$$

$$\times \left\{ \left\| P^{(k)} \right\| \left\| x^{(k)} - x^{(k-1)} \right\| + \left\| P^{(k)} - P^{(k-1)} \right\| \left\| x^{(k-1)} \right\| \right\}$$

$$\left\|\widetilde{S}^{(k)} - \widetilde{S}^{(k-1)}\right\| \le \left[\sum_{j=1}^{n} \left\|\frac{1}{2}(N_j R^{-1} B^T + B R^{-1} N_j^T)\right\|^2\right]^{\frac{1}{2}}$$

$$\times \left\|x^{(k)} - x^{(k-1)}\right\|$$

$$+ \left[\sum_{i,j=1}^{n} \left\|N_i R^{-1} N_j^T\right\|^2\right]^{\frac{1}{2}} \left(\left\|x^{(k)}\right\| + \left\|x^{(k-1)}\right\|\right) \left\|x^{(k)} - x^{(k-1)}\right\|$$

$$(4.67)$$

$$\left\|\widetilde{Q}^{(k)} - \widetilde{Q}^{(k-1)}\right\| \le \left[\sum_{i,j=1}^{n} \left\|\frac{1}{2}(N_i R^{-1} N_j^T + N_j R^{-1} N_i^T)\right\|^2\right]^{\frac{1}{2}}$$

$$(4.68)$$

$$\times \left(\left\|P^{(k)} x^{(k)}\right\| + \left\|P^{(k-1)} x^{(k-1)}\right\|\right)$$

$$\times \left\{\left\|P^{(k)}\right\|\left\|x^{(k)} - x^{(k-1)}\right\| + \left\|P^{(k)} - P^{(k-1)}\right\|\left\|x^{(k-1)}\right\|\right\}$$

Application of the results of (4.64)-(4.68) to (4.62) leads to the same fixed-point problem as the one obtained in (Hofer and Tibken, 1988)

$$\begin{bmatrix} \left\|x^{(k+1)} - x^{(k)}\right\| \\ \left\|P^{(k+1)} - P^{(k)}\right\| \end{bmatrix} \le M \begin{bmatrix} \left\|x^{(k)} - x^{(k-1)}\right\| \\ \left\|P^{(k)} - P^{(k-1)}\right\| \end{bmatrix} \qquad (4.69)$$

where the 2×2 matrix M is given by

$$M = \begin{bmatrix} \mu_1 & \mu_2 \\ \mu_3 & \mu_4 \end{bmatrix} \|R^{-1}\| \qquad (4.70)$$

The rest of the convergence proof follows by invoking Theorem 4.1 from (Hofer and Tibken, 1988) which states the contraction property for a pair of operators defined in (4.69) under the assumption that the eigenvalues of the matrix M are inside of the unit circle. It is important to notice that the multiplicative influence of R^{-1} in (4.70) makes the eigenvalues of the matrix M arbitrarily small by choosing R arbitrarily large.

4.6 Case Study: Chemical Reactor

The new method for the optimal control of bilinear systems is applied to the control of a chemical reactor, (Hofer and Tibken, 1988). The bilinear model of the system is given by

$$A = \begin{bmatrix} 13/6 & 5/12 \\ -50/3 & -8/3 \end{bmatrix}, \qquad B = \begin{bmatrix} -1/8 \\ 0 \end{bmatrix}$$

$$N_1 = \begin{bmatrix} -1 \\ 0 \end{bmatrix}, \qquad N_2 = \begin{bmatrix} 0 \\ 0 \end{bmatrix}$$

The normalized dimensionless state variables x_1 and x_2 represent temperature and concentration of the initial product of the chemical reaction, respectively. The normalized dimensionless control u represents the cooling flow rate in a jacket around the reactor. In order to transfer the system in finite-time very closely to the steady state given by $x = 0$, $u = 0$, the weighting matrix F in the performance index has to be chosen dominant compared to the design matrices Q and R. A choice of the design matrices F, Q, and R is

$$F = \begin{bmatrix} 1000 & 0 \\ 0 & 1000 \end{bmatrix}, \qquad Q = \begin{bmatrix} 10 & 0 \\ 0 & 10 \end{bmatrix}, \qquad R = 1$$

and the initial conditions are $x_0 = \begin{bmatrix} 0.15 & 0 \end{bmatrix}^T$. Simulation results are presented in Figures 4.8 and 4.9 where the solid lines are the optimal trajectories, the dashed lines are the first approximations, the dotted lines are the second approximations, and the dashed-dotted lines are the third approximations.

It can be seen from Figures 4.8 and 4.9 that the new method preserves, in this particular example, very good convergence. In addition, the convergence of Algorithm 4.3 is achieved with relatively small value for the control penalty matrix R so that the constraint imposed in Assumption 4.4 does not seem to be very severe.

All numerical results in this book are obtained by using MATLAB software and its control tool box.

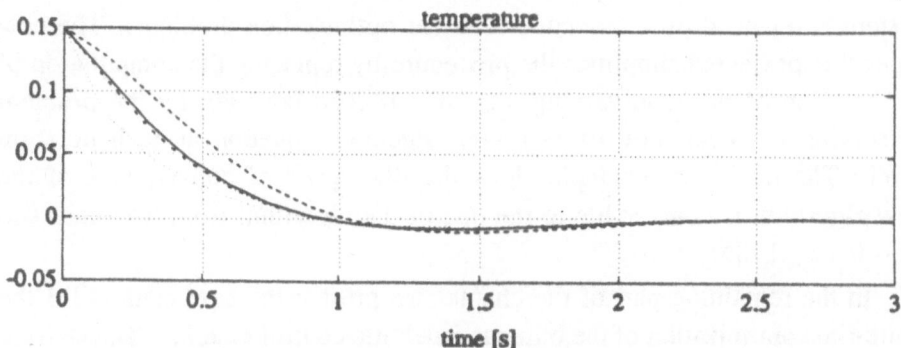

Figure 4.8: Profiles of temperature for $x^0 = (0.15, \ 0)^T$

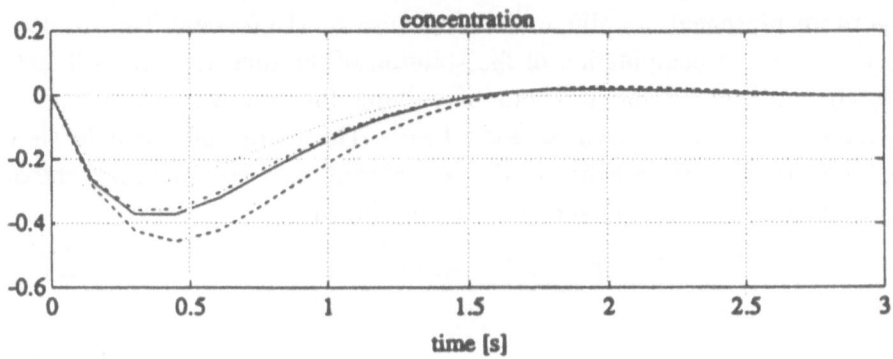

Figure 4.9: Profiles of concentration for $x^0 = (0.15, \ 0)^T$

4.7 Conclusion

In this chapter, the new method for the steady state optimization of the bilinear-quadratic control problem is presented. The starting point is the algorithm of (Cebuhar and Constanza, 1984) for the approximation of the optimal solution of the same problem. The method of (Cebuhar and Constanza, 1984) itself presents an interesting approach from the application point of view. Namely, the optimization problem of a bilinear (nonlinear) system is replaced by a sequence of linear optimization problems. The new algorithm presented simplifies the procedure by replacing the computation of the solution of the time-varying algebraic Riccati equation by the problem of solving the Lyapunov time-varying algebraic equation at each iteration level. The numerical example show that the speed of convergence of the new algorithm is comparable to the one of the algorithm from (Cebuhar and Constanza, 1984).

In the remaining part of the chapter we present the new method for the finite-time optimization of the bilinear-quadratic control systems. The starting point is the algorithm of (Hofer and Tibken, 1988) for the approximation of the optimal solution of the bilinear optimal control problem. That method itself presents an interesting approach from the application point of view. Namely, the optimization problem of the bilinear (nonlinear) system is replaced by a sequence of the linear optimization problems. The new algorithm presented simplifies the procedure of (Hofer and Tibken, 1988) by replacing the computation of the solution of the time-varying differential Riccati equation by the problem of solving the time-varying differential Lyapunov equation at each iteration level. The numerical example shows that the speed of convergence of the new algorithm is not inferior to the one of the algorithm from (Hofer and Tibken, 1988).

Appendix 4.1

Successive Approximations for Steady State Linear-Quadratic Optimal Control Problem

Consider the linear-quadratic optimal control problem

$$\dot{x}(t) = A(t)x(t) + B(t)u(t) \tag{a.1}$$

$$J = \frac{1}{2} \int_t^\infty \left[x^T(\tau)Q(\tau)x(\tau) + u^T(\tau)R(\tau)u(\tau) \right] d\tau \tag{a.2}$$

Note that the optimization problem defined in (a.1)-(a.2) is more general than the one with fixed initial time, (Kirk, 1970). Corresponding Hamiltonian is given by

$$H\left(x, u, \frac{\partial J}{\partial x}, t\right) = \left\{ x^T(t)Q(t)x(t) + u^T(t)R(t)u(t) \right\}$$
$$+ \left(\frac{\partial J}{\partial x}(t) \right)^T \left\{ A(t)x(t) + B(t)u(t) \right\} \tag{a.3}$$

The successive approximations technique applied to (a.1)-(a.3) is composed of the following steps.

Step 1. Take any stabilizable linear control law $u^{(0)}(x(t))$, for example $u^{(0)}(x(t)) = -R^{-1}(t)B^T(t)P^{(0)}(t)x(t)$ with $P^{(0)}(t)$ being symmetric, and find the expression for $\frac{\partial J^{(0)}}{\partial x}(t)$ of the performance criterion

$$J^{(0)} = \frac{1}{2} \int_t^\infty \left[x^T(\tau)Q(\tau)x(\tau) + u^{(0)^T}(x(\tau))R(\tau)u^{(0)}(x(\tau)) \right] d\tau$$
$$= \frac{1}{2} \int_t^\infty x^T(\tau) \left[Q(\tau) + P^{(0)}(\tau)B^T(\tau)R(\tau)B(\tau)P^{(0)}(\tau) \right] x(\tau) d\tau \tag{a.4}$$

along the trajectories of the system

$$\dot{x}(t) = A(t)x(t) + B(t)u^{(0)}(x(t)) = \left[A(t) - S(t)P^{(0)}(t) \right] x(t)$$
$$S(t) = B^T(t)R^{-1}(t)B(t) \tag{a.5}$$

Step 2. For the known value of $\frac{\partial J^{(0)}}{\partial x}(t)$ find a new approximation for the control law by minimizing with respect to u the "partially frozen" Hamiltonian

$$H\left(x, u, \left(\frac{\partial J}{\partial x}\right)^{(0)}, t\right) = (x^T(t)Q(t)x(t) + u^T(t)R(t)u(t))$$
$$+ \left(\frac{\partial J}{\partial x}(t)\right)^{(0)^T} (A(t)x(t) + B(t)u(t)) \tag{a.6}$$

The minimization produces a stabilizing control given by

$$u^{(1)}(t) = -R^{-1}(t)B^T(t)\left(\frac{\partial J}{\partial x}(t)\right)^{(0)} \tag{a.7}$$

Note that $\frac{\partial J}{\partial x}$ can be calculated from (a.1)-(a.2), namely by using the identity

$$\frac{dJ}{dt} = \frac{\partial J}{\partial x}\frac{dx}{dt} = -\frac{1}{2}(x^T(t)Q(t)x(t) + u^T(t)R(t)u(t)) \tag{a.8}$$

Under the stabilizing control $u^{(0)}(x(t))$ the last equality produces

$$\frac{\partial J^{(0)}}{\partial x}\left(A(t) - S(t)P^{(0)}(t)\right)x(t) = -\frac{1}{2}x^T(t)\left[Q(t) + P^{(0)}(t)S(t)P^{(0)}(t)\right] \tag{a.9}$$

This simple partial differential equation has a solution of the form

$$J^{(0)} = \frac{1}{2}x^T(t)P^{(1)}(t)x(t) \tag{a.10}$$

By using the fact that

$$\frac{\partial J^{(0)}}{\partial x} = P^{(1)}(t)x(t) \tag{a.11}$$

we get

$$x^T(t)P^{(1)}(t)\left(A(t) - S(t)P^{(0)}(t)\right)x(t)$$
$$= -\frac{1}{2}x^T(t)\left[Q(t) + P^{(0)}(t)S(t)P^{(0)}(t)\right] \tag{a.12}$$

Using the standard symmetrization technique known from the derivations of the Riccati equation, that is

$$x^T M x = \frac{1}{2} x^T (M + M^T) x, \quad for \ any \ square \ matrix \ M \quad \text{(a.13)}$$

we get

$$\left(A(t) - S(t) P^{(0)}(t) \right)^T P^{(1)}(t) + P^{(1)}(t) \left(A(t) - S(t) P^{(0)}(t) \right)$$
$$+ \left(Q(t) + P^{(0)}(t) S(t) P^{(0)}(t) \right) = 0 \quad \text{(a.14)}$$

Then

$$\frac{\partial J^{(0)}}{\partial x}(t) = P^{(1)}(t) x(t) \Rightarrow u^{(1)}(t) = -R^{-1}(t) B^T(t) P^{(1)}(t) x(t) \quad \text{(a.15)}$$

The monotonicity result of successive approximations states, (Bellman, 1961; Leake and Liu, 1967; Mil'shtein, 1964)

$$J^{(0)}(t) \geq J^{(1)}(t) \Leftrightarrow P^{(0)}(t) \geq P^{(1)}(t) \quad \text{(a.16)}$$

By repeating Steps 1 and 2 now with $u^{(1)}(x(t))$ we get $u^{(2)}(x(t))$ and $P^{(2)}(t)$. Continuing the same procedure we get a monotonically decreasing sequence

$$P^{(0)}(t) \geq P^{(1)}(t) \geq P^{(2)}(t) \geq \quad ..... \quad \text{(a.17)}$$

This sequence is also convergent since $P^{(m)}(t)$, $m = 0, 1, 2, ..$ are positive operators, (Kleinman, 1968; Kantorovich and Akilov, 1964).

Appendix 4.2

Successive Approximations for Finite-Time
Linear-Quadratic Optimal Control Problem

Consider the time-varying finite-time linear-quadratic optimal control problem

$$\dot{x}(t) = A(t)x(t) + B(t)u(t) \tag{b.1}$$

$$J = \frac{1}{2}x^T(t_f)Fx(t_f) + \frac{1}{2}\int_t^{t_f} \left[x^T(\tau)Q(\tau)x(\tau) + u^T(\tau)R(\tau)u(\tau)\right]d\tau \tag{b.2}$$

Note that the optimization problem defined in (b.1)-(b.2) is more general than the one with fixed initial time, (Kirk, 1970). Corresponding Hamiltonian is given by

$$H\left(x, u, \frac{\partial J}{\partial x}, t\right) = \left\{x^T(t)Q(t)x(t) + u^T(t)R(t)u(t)\right\}$$
$$+ \left(\frac{\partial J}{\partial x}(t)\right)^T \left\{A(t)x(t) + B(t)u(t)\right\} \tag{b.3}$$

The successive approximations technique applied to (b.1)-(b.3) is composed of the following steps.

Step 1. Take any stabilizable linear control law $u^{(0)}(x(t))$, for example $u^{(0)}(x(t)) = -R^{-1}(t)B^T(t)P^{(0)}(t)x(t)$ with $P^{(0)}(t)$ being symmetric, and find the expression for $\frac{\partial J^{(0)}}{\partial x}(t)$ of the performance criterion

$$J^{(0)} = \frac{1}{2}x^T(t_f)Fx(t_f)$$

$$+ \frac{1}{2}\int_t^{t_f} \left[x^T(\tau)Q(\tau)x(\tau) + u^{(0)^T}(x(\tau))R(\tau)u^{(0)}(x(\tau))\right]d\tau$$

$$= \frac{1}{2}x^T(t_f)Fx(t_f) \tag{b.4}$$

$$+ \frac{1}{2}\int_t^{t_f} x^T(\tau)\left[Q(\tau) + P^{(0)}(\tau)B^T(\tau)R(\tau)B(\tau)P^{(0)}(\tau)\right]x(\tau)d\tau$$

along the trajectories of the system

$$\dot{x}(t) = A(t)x(t) + B(t)u^{(0)}(x(t)) = \left[A(t) - S(t)P^{(0)}(t)\right]x(t)$$
$$S(t) = B^T(t)R^{-1}(t)B(t)$$

(b.5)

Step 2. For the known value of $\frac{\partial J^{(0)}}{\partial x}(t)$ find a new approximation for the control law by minimizing with respect to u the "partially frozen" Hamiltonian

$$H\left(x, u, \left(\frac{\partial J}{\partial x}\right)^{(0)}, t\right) = (x^T(t)Q(t)x(t) + u^T(t)R(t)u(t))$$
$$+\left(\frac{\partial J}{\partial x}(t)\right)^{(0)^T}(A(t)x(t) + B(t)u(t))$$

(b.6)

The minimization produces a stabilizing control given by

$$u^{(1)}(t) = -R^{-1}(t)B^T(t)\left(\frac{\partial J}{\partial x}(t)\right)^{(0)}$$

(b.7)

Note that $\frac{\partial J^{(0)}}{\partial x}(t)$ can be obtained from (b.4)-(b.5) by using known results from (Kwakernaak and Sivan, 1972) as

$$J^{(0)} = \frac{1}{2}x^T(t)P^{(1)}(t)x(t)$$

(b.8)

where

$$-\dot{P}^{(1)}(t) = \left(A(t) - S(t)P^{(0)}(t)\right)^T P^{(1)}(t)$$
$$+P^{(1)}(t)\left(A(t) - S(t)P^{(0)}(t)\right) + \left(Q(t) + P^{(0)}(t)S(t)P^{(0)}(t)\right)$$
$$P(t_f) = F$$

(b.9)

Then

$$\frac{\partial J^{(0)}}{\partial x}(t) = P^{(1)}(t)x(t) \Rightarrow u^{(1)}(t) = -R^{-1}(t)B^T(t)P^{(1)}(t)x(t)$$

(b.10)

The monotonicity result of successive approximations states, (Bellman, 1961; Leake and Liu, 1967; Milshtein, 1964)

$$J^{(0)}(t) \geq J^{(1)}(t) \Leftrightarrow P^{(0)}(t) \geq P^{(1)}(t) \tag{b.11}$$

By repeating steps 1 and 2 now with $u^{(1)}(x(t))$ we get $u^{(2)}(x(t))$ and $P^{(2)}(t)$. Continuing the same procedure we get a monotonically decreasing sequence

$$P^{(0)}(t) \geq P^{(1)}(t) \geq P^{(2)}(t) \geq \quad ..... \tag{b.12}$$

This sequence is also convergent since $P^{(m)}(t)$, $m = 0, 1, 2, ..$ are positive operators, (Kantorovich and Akilov, 1964; Kleinman 1968).

Chapter 5

Concluding Remarks

In this chapter we propose and outline some of the future research problems on singularly perturbed and weakly coupled bilinear systems with emphasis on linear optimal control in the *discrete-time* domain. We also indicate the importance of studying *stochastic* singularly perturbed and weakly coupled bilinear systems.

In the previous chapters we have presented results for linear optimal control of *continuous-time* bilinear singularly perturbed and weakly coupled systems. As the first research problem we suggest to study the linear optimal control of singularly perturbed and weakly coupled bilinear continuous-time systems by the means of the methodology presented in Chapter 4. In that respect the Riccati equations will be replaced by the Lyapunov equations, which will simplify the actual computations and speed up the convergence of the numerical iterative techniques used to solve these optimal control problems.

We have pointed out before the importance of extending the results obtained in Chapters 2 and 3 to discrete-time bilinear singularly perturbed and weakly coupled systems. The theory of optimal control of *linear* discrete-time systems (Dorato and Levis, 1970; Sage and White, 1977; Lewis, 1986) is presently pretty much developed also for both singularly perturbed and weakly coupled systems (Mahmoud, 1982; Litkouhi and Khalil, 1984, 1985;

Shen and Gajic, 1990b, 1990c; Gajic and Shen, 1991b; Aganovic et al., 1994; Lim et al., 1995).

Discrete-time bilinear control systems are represented by

$$x(k+1) = Ax(k) + \sum_{i=1}^{m} u_i(k)N_i x(k) + Bu(k)$$

$$\text{(5.1)}$$

$$= Ax(k) + \{u(k)N\}x(k) + Bu(k)$$

where $x(k) \in \Re^n$ are state variables, $u(k) \in \Re^m$ are control inputs and A, B, N_i are constant (or time-varying) matrices of compatible dimensions, and k represents the discrete-time instants. The corresponding quadratic performance criterion to be minimized is given by

$$J = \sum_{k=0}^{\infty} \left[x^T(k)Qx(k) + u^T(k)Ru(k) \right], \quad Q \ge 0, \ R > 0 \quad \text{(5.2)}$$

Formula (5.2) reflects the steady state optimization. Similarly, we can define the quadratic performance criterion for the finite time optimization problem by using a finite sum in (5.2).

The general optimization problem of *discrete-time bilinear systems* with a quadratic performance criterion should be studied for both the finite-time and infinite-time (steady state) optimization horizons. In that direction one has first to extend the results of (Bruni et al, 1971; Cebuhar and Constanza, 1984; Hofer and Tibken, 1988) to the discrete-time domain. For the obtained discrete-time results one may consider the singularly perturbed and weakly coupled bilinear system structures. In addition, after the successful extension of the work of (Bruni et al, 1971; Cebuhar and Constanza, 1984; Hofer and Tibken, 1988) to the discrete-time domain, the simplification of the results obtained in terms of the discrete-time algebraic and difference Riccati equations may be attempted along the line of results of (Aganovic and Gajic, 1994, 1995) as presented in Chapter 4 of this book. This might be achieved by exploiting the power of the successive approximations technique of dynamic programming. Thus, we expect that the bilinear-quadratic optimal control problem will be solved by using the linear controllers whose matrix coefficients are obtained from sequences of discrete-time Lyapunov equations (difference and/or algebraic ones). Finally, having obtained all of the above results one may again consider the linear optimal control for the special structures of singularly perturbed and weakly coupled discrete-time bilinear systems.

Stochastic bilinear control systems have been studied by several authors, see for example (Swamy and Tran, 1979; Mohler and Kolodziej, 1981; Kubrusly and Costa, 1985; Desai, 1986; Yaz, 1992) and references therein. The bilinear discrete-time stochastic systems are represented by

$$x(k + 1) = Ax(k) + \{u(k)N\}x(k) + Bu(k) + Gw(k)$$

$$(5.3)$$

$$y(k) = Cx(k) + v(k)$$

where $y(k) \in \Re^r$ is the system measurement vector, $w(k) \in \Re^s$ and $v(k) \in \Re^r$ are stochastic disturbances in most cases considered as zero-mean Gaussian white noise stochastic processes. Study of stochastic singularly perturbed and weakly coupled bilinear systems may be an interesting and very challenging topic for future research. In that direction both continuous-time and discrete-time stochastic problems should be considered. We want to point out that several results are already obtained in the context of singular perturbations and weak coupling for the *linear* continuous-time and discrete-time stochastic systems (Haddad, 1977; Haddad and Kokotovic, 1977; Khalil and Gajic, 1984; Gajic, 1986; Gajic and Khalil, 1986; Shen and Gajic, 1990c; Gajic and Shen, 1991a; Gajic and Lim, 1994; Aganovic et al., 1994; Gajic and Aganovic, 1995; Lim et al., 1995).

A special class of stochastic bilinear control systems are *linear systems with state- and control-dependent disturbances* (Wonham, 1967; Kleinman, 1969; McLane, 1971). In these systems the bilinear term comes either from the product of state variables and noise (disturbance) and/or the product of the control variable and noise. The complete results have been obtained in the above papers for the case of Gaussian white noise state- and control-dependent disturbances. The optimal controllers for this special class of bilinear systems have been obtained in terms of the solutions of the Riccati-like equations. Extension of these results to the singularly perturbed and weakly coupled systems with state- and/or control-dependent disturbances may be an interesting and important subject for future research.

We would like to emphasize the power of the results obtained by (Su et al, 1992b; Gajic and Lim, 1994; Aganovic et al., 1994; Gajic and Aganovic, 1995; Lim et al., 1995) for the decomposition of the optimal control and filtering tasks of linear singularly perturbed and weakly coupled systems. In these papers the optimal filtering and control are obtained by using the *closed-loop* decomposition technique such that both optimal filtering and control are independently and exactly solved in terms of the reduced-order optimal filtering and control subproblems, which are combined to get the full order global optimal solutions. Incorporation of these results into bilinear singularly perturbed and weakly coupled filtering and control problems might produce tremendous savings of required computations and exact and full parallelism of these tasks at the subsystem levels.

In conclusion, we would like to emphasize that the bilinear singularly perturbed and weakly coupled control systems are very fruitful areas for future research. Only a few results have been obtained so far. In addition, the bilinear control systems as a part of the much large group of nonlinear control systems are the research imperative of modern time (second part of the eighties and the first part of the nineties).

List of References

1. **Aganovic**, Z. (1993). *Singularly Perturbed and Weakly Coupled Bilinear Control Systems*. Ph. D. Dissertation, Rutgers University.

2. Aganovic, Z. and Z. Gajic, (1993). Optimal control of weakly coupled bilinear systems. *Automatica*, **29**, 1591–1593.

3. Aganovic, Z. and Z. Gajic, (1994). The successive approximation procedure for finite-time optimal control of bilinear systems. *IEEE Trans. Automatic Control*, **AC-39**, 1932–1935.

4. Aganovic, Z. and Z. Gajic, (1995). The successive approximation procedure for steady state optimal control of bilinear systems. *J. Optimization Theory and Applications*, **84**, 273–291.

5. Aganovic, Z., Z. Gajic, and X. Shen, (1995). New method for optimal control and filtering of weakly coupled discrete stochastic linear systems. *Proc. Conference on Decision and Control*, 1–6, Orlando, FL.

6. Aldhaheri, R. and H. Khalil, (1991). Aggregation of the policy iteration method for nearly completely decomposable Markov chains. *IEEE Trans. Automatic Control*, **AC-36**, 178–187.

7. Arabacioglu, M., M. Sezer, and O. Oral, (1986). Overlapping decomposition of large scale systems into weakly coupled subsystems, 135–147, in *Computational and Combinatorial Methods in System Theory*, C. Byrnes and A. Lindquist, eds., North Holland, Amsterdam.

8. Asamoah, F. and M. Jamshidi, (1987). Stabilization of a class of singularly perturbed bilinear systems. *Int. J. Control*, **46**, 1589–1594.

9. **Baheti**, R. and R. Mohler, (1981). Experimental results in the modeling and control of a small furnace. *ASME J. Dyn. Sys. Meas. and Control*, **103**, 370–374.

10. Bahrami, K. and M. Kim, (1975). Optimal control of multiplicative control systems arising from cancer therapy. *IEEE Trans. Automatic Control*, **AC-20**, 537–541, 1975.

11. Balakrishnan, A., (1976). Are all nonlinear systems bilinear. *Proc. Joint American Control Conference*.

12. Banks, S. and M. Yew, (1985). On a class of suboptimal controls for infinite-dimensional bilinear systems. *Systems & Control Letters*, **5**, 327–333.

13. Banks, S. and M. Yew, (1986). On optimal control of bilinear systems and its relation to Lie algebras. *Int. J. Control*, **43**, 891–900.

14. Baruh, H. and K. Choe, (1990). Sensor placement in structural control. *AIAA J. Guidance, Dynamics and Control*, **13**, 524–533.

15. Bellman, R., (1954). Monotone approximation in dynamic programming and calculus of variations, *Proc. The National Academy of Science, USA*, **44**, 1073–1075.

16. Bellman, R., (1957). *Dynamic Programming*, Princeton University Press, Princeton.

17. Bellman, R., (1961). *Adaptive Control Processes: A Guided Tour*, Princeton University Press, Princeton.

18. Benallou, A., D. Mellichamp, and A. Seborg, (1988). Optimal stabilizing controllers for bilinear systems, *Int. J. Control*, **48**, 1487–1501.

19. Biran, Y. and B. McInnus, (1979). Optimal control of bilinear systems, time-varying effects of cancer drugs. *Automatica*, **15**, 325–329.

20. Borno, I. and Z. Gajic, (1995). Parallel algorithms for optimal control of weakly coupled and singularly perturbed jump linear systems. *Automatica*, **31**, in press.

21. Bruni, C., G. DiPillo, and G. Koch, (1971). On the mathematical models of bilinear systems, *Ricerche di Automatica*, **2**, 11–26.

22. Bruni, C., G. DiPillo, and G. Koch, (1974). Bilinear systems: an appealing class of "nearly linear" systems in theory and applications. *IEEE Trans. Automatic Control*, **AC-19**, 334–348.

23. **Cebuhar**, W. and V. Constanza, (1984). Approximation procedures for the optimal control of bilinear and nonlinear systems. *J. of Optimization Theory and Applications*, **43**, 615–627.

24. Chang, K. (1972). Singular perturbations of a general boundary value problem. *SIAM J. Math. Anal.*, **3**, 520–526.

25. Chen, L., X. Yang, and R. Mohler, (1991). Stability analysis of bilinear systems. *IEEE Trans. Automatic Control*, **AC-36**, 1310–1315.

26. Chow, J. and P. Kokotovic, (1976). A decomposition of near-optimum regulators for systems with slow and fast modes. *IEEE Trans. Automatic Control*, **AC-21**, 701–706.

27. Chow J. and P. Kokotovic, (1978a). Two-time scale feedback design of a class of nonlinear systems. *IEEE Trans. Automatic Control*, **AC-23**, 438-443.

28. Chow, J. and P. Kokotovic, (1978b). Near-optimal feedback stabilization of a class of nonlinear singularly perturbed systems. *SIAM J. Control and Optimization*, **16**, 756–770.

29. Chow, J. and P. Kokotovic, (1981). A two-stage Lyapunov-Bellman feedback design of a class of nonlinear systems. *IEEE Trans. Automatic Control*, **AC-26**, 656–663.

30. Chow, J. and P. Kokotovic, (1983). Sparsity and time scales. *Proc. American Control Conference*, 656–661, San Francisco, CA.

31. Cronin, J., (1987). *Mathematical Aspects of Hodgkin-Huxley Neural Theory*, Cambridge Press, 1987.

32. **Delacour**, J., M. Darwish, and J. Fantin, (1978). Control strategies of large-scale power systems. *Int. J. Control*, **27**, 753–767.

33. Delebecque, F. and J. Quadrat, (1981). Optimal control of Markov chains admitting strong and weak interconnections. *Automatica*, **17**, 281–296.

34. Derese, I. and E. Noldus, (1980). Design of linear feedback laws for bilinear systems, *Int. J. Control*, **31**, 219–237.

35. Desai, U., (1986). Realization of bilinear stochastic systems. *IEEE Trans. Automatic Control*, **AC-31**, 189–192.

36. Desoer, C. and M. Shena, (1970). Networks with very small and very large parasitics: natural frequencies and stability. *Proc. IEEE*, **58**, 1933–1938.

37. Dorato, P. and A. Levis, (1970). Optimal linear regulators: the discrete time case. *IEEE Trans. Automatic Control*, **AC-16**, 613–620.

38. **Espana**, M. and I. Landau, (1978). Reduced order bilinear models for distillation columns. *Automatica*, **14**, 345–355.

39. **Figalli**, G., M. Cava, and L. Tomasi, (1984). An optimal feedback control for a bilinear model of induction motor drives. *Int. J. Control*, **39**, 1007–1016.

40. **Gajic**, Z., (1986). Numerical fixed-point solution for near-optimum regulators of linear quadratic Gaussian control problems for singularly perturbed systems. *Int. J. Control*, **43**, 373–387.

41. Gajic, Z., (1988). Existence of a unique and bounded solution of the algebraic Riccati equation of the multimodel estimation and control problems. *Systems & Control Letters*, **10**, 185–190.

42. Gajic, Z. and H. Khalil, (1986). Multimodel strategies under random disturbances and imperfect partial observations. *Automatica*, **22**, 121–125.

43. Gajic, Z. and M. Lim, (1994). A new filtering method for linear singularly perturbed systems. *IEEE Trans. Automatic Control*, **AC-38**, 1952–1955.

44. Gajic, Z., D. Petkovski, and N. Harkara, (1989). The recursive algorithm for the optimal static output feedback control problem of linear singularly perturbed systems. *IEEE Trans. Automatic Control*, **AC-34**, 465–468.

45. Gajic, Z., D. Petkovski, and X. Shen, (1990). *Singularly Perturbed and Weakly Coupled Linear Control Systems — A Recursive Approach*, Springer-Verlag, New York.

46. Gajic, Z. and M. Qureshi, (1995). *The Lyapunov Matrix Equation in System Stability and Control*, Academic Press, Boston, 1995.

47. Gajic, Z., and X. Shen, (1989). Decoupling transformation for weakly coupled linear systems. *Int. J. Control*, **50**, 1515–1521.

48. Gajic, Z. and X. Shen, (1991a). Parallel reduced-order controllers for stochastic linear singularly perturbed discrete systems. *IEEE Trans. Automatic Control*, **AC-35**, 87–90.

49. Gajic, Z. and X. Shen, (1991b). Study of the discrete singularly perturbed linear-quadratic control problem by a bilinear transformation. *Automatica*, **27**, 1025–1028.

50. Gajic, Z. and X. Shen, (1993). *Parallel Algorithms for Optimal Control of Large Scale Linear Systems*, Springer-Verlag, London.

51. Gajic, Z. and Z. Aganovic, (1995). New filtering method for linear weakly coupled stochastic systems. *AIAA J. Guidance, Control and Dynamics*, **18**, in press.

52. Guillen, J. and M. Armada, (1980). Singular perturbation method for order reduction of large-scale bilinear dynamical systems. *Proc. IFAC Symp. Large-Scale Systems Theory and Appl.*, 229–236, Touluse.

53. Guo, L., A. Schone, and X. Ding, (1994). Control of hydraulic rotary multi-motor systems based on bilinearization. *Automatica*, **30**, 1445–1453.

54. Gutman, P. (1981). Stabilizing controls for bilinear systems. *IEEE Trans. Automatic Control*, **AC-26**, 917–922, 1981.

55. **Haddad**, A., (1976). Linear filtering of singularly perturbed systems. *IEEE Trans. Automatic Control*, **AC-21**, 515–519.

56. Haddad, A. and P. Kokotovic, (1977). Stochastic control of linear singularly perturbed systems. *IEEE Trans. Automatic Control*, **AC-22**, 815–821.

57. Harkara, N., D. Petkovski, and Z. Gajic, (1989). The recursive algorithm for optimal output feedback control problem of linear weakly coupled systems. *Int. J. Control*, **50**, 1–11.

58. Hill, D. *Experiments in Computational Matrix Algebra*. Random House, New York, 1988.

59. Hofer, E. and B. Tibken, (1988). An iterative method for the finite-time bilinear quadratic control problem. *J. Optimization Theory and Applications*, **57**, 411–427.

60. **Ikeda**, M. and D. Siljak, (1980). Overlapping decompositions expansions and contractions of dynamic systems. *Large Scale Systems*, **1**, 29–38.

61. Ionescu, T. and R. Monopoli, (1975). Stabilization of bilinear systems via hyperstability. *IEEE Trans. Automatic Control*, **AC-20**, 280–284.

62. Ishimatsu, T., A. Mohri, and M. Takata. (1975). Optimization of weakly coupled systems by a two-level method. *Int. J. Control*, **22**, 877–882.

63. **Jacobson**, D., (1980). *Extensions of Linear-Quadratic Control Systems*, Springer Verlag, Berlin.

64. **Kalman**, R., (1960). Contributions to the theory of optimal control. *Bol. Soc. Mat. Mex.*, 102–119.

65. Kantorovich, L. and G. Akilov, (1964). *Functional Analysis in Normed Spaces*, Macmillan, New York.

66. Kaszkurewicz, E., A. Bhaya, and D. Siljak, (1990). On the convergence of parallel asynchronous block-iterative computations. *Linear Algebra and Its Applications*, **131**, 139–160.

67. Khalil, H., (1980). Multi-model design of a Nash strategy. *J. Optimization Theory and Applications*, **31**, 553–564.

68. Khalil, H., (1992). *Nonlinear Systems*. Macmillan, 1992.

69. Khalil, P. and P. Kokotovic, (1978). Control strategies for decision makers using different models of the same system. *IEEE Trans. Automatic Control*, **AC-23**, 289–298.

70. Khalil, P. and Z. Gajic, (1984). Near-optimum regulators for stochastic linear singularly perturbed systems. *IEEE Trans. Automatic Control*, **AC-29**, 531–541.

71. Kirk, D. (1970). *Optimal Control Theory*, Prentice-Hall, Englewood Cliffs.

72. Kleinman, D., (1968). On iterative technique for Riccati equation computations, *IEEE Trans. Automatic Control*, **AC-13**, 114–115.

73. Kleinman, D., (1969). Optimal stationary control of linear systems with control dependent noise, *IEEE Trans. Automatic Control*, **AC-14**, 673–677.

74. Kleinman, D., (1970). An easy way to stabilize a linear constant system, *IEEE Trans. Automatic Control*, **AC-15**, 692.

75. Ko, K., B. Mcinnis, and G. Goodwin, (1982). Adaptive control and identification of the dissolved oxygen process. *Automatica*, **18**, 727–730.

76. Kokotovic, P., W. Perkins, J. Cruz, and G. D'Ans (1969). ϵ–coupling for near-optimum design of large scale linear systems. *Proc. IEE, Part D.*, **116**, 889–892.

77. Kokotovic, P. and G. Singh, (1971). Optimization of coupled nonlinear systems. *Int. J. Control*, **14**, 51–64.

78. Kokotovic, P. and R. Yackel, (1972). Singular perturbation of linear regulators: Basic theorems. *IEEE Trans. Automatic Control*, **AC-17**, 29–37.

79. Kokotovic, P., H. Khalil, and J. O'Reilly, (1986). *Singular Perturbation Methods in Control: Analysis and Design*, Academic Press, Orlando.

80. Kokotovic, P. and H. Khalil, Eds., (1986). *Singular Perturbation in Systems and Control*, IEEE Press, New York.

81. Kubrusly, C. and O. Costa, (1985). Mean square stability conditions for discrete stochastic bilinear systems. *IEEE Trans. Automatic Control*, **AC-30**, 1082–1087.

82. Kucera, V. (1972). A contribution to matrix quadratic equations. *IEEE Trans. Automatic Control*, **AC-17**, 344–347.

83. Kwakernaak, H. and R. Sivan, (1972). *Linear Optimal Control Systems*, Wiley, New York.

84. **Larson**, R., (1967). A Survey of dynamic programming computational procedures, *IEEE Trans. Aut. Control*, **AC-12**, 767–774.

85. Leake, R. and R. Liu, (1967). Construction of suboptimal control sequences, *SIAM J. Control*, **5**, 54–63.

86. Levine, M. and T. Vilis, (1973). On-line learning optimal control using successive approximation techniques, *IEEE Trans. Automatic Control*, **AC-19**, 279–284.

87. Lewis, F. *Optimal Control*. Wiley, New York, 1986.

88. Lim, M., Z. Gajic, and X. Shen, (1995). New methods for optimal control and filtering of singularly perturbed linear discrete stochastic systems. *Proc. American Control Conference*, Seattle, WA, in press.

89. Litkouhi, B. and H. Khalil, (1984). Infinite-time regulators for singularly perturbed difference equations. *Int. J. Control*, **39**, 587–598.

90. Litkouhi, B. and H. Khalil, (1985). Multirate and composite control of two-time-scale discrete systems. *IEEE Trans. Aut. Control*, **AC-30**, 645–651.

91. Lo, J., (1975). Global stabilization with control appearing linearly. *SIAM J. Control*, **13**, 875–885.

92. Longchamp, R., (1980a). State feedback control of bilinear systems. *IEEE Trans. Automatic Control*, **AC-25**, 302–306.

93. Longchamp, R., (1980b). Controller design for bilinear systems. *IEEE Trans. Automatic Control*, **AC-25**, 547–548.

94. **Mageriou**, E., (1977). Iterative techniques for Riccati game equations, *J. Optimization Theory and Applications*, **22**, 51–61.

95. Mahmoud, M. (1978). A quantitative comparison between two decentralized control approaches. *Int. J. Control*, **28**, 261–275.

96. Mahmoud, M. (1982). Order reduction and control of discrete systems. *Proc. IEE, Part D.*, **129**, 129–135.

97. McLane, P., (1971). Optimal stochastic control of linear systems with state- and control-dependent disturbances. *IEEE Trans. Automatic Control*, **AC–16**, 793–798.

98. Medanic, J. and B. Avramovic, (1975). Solution of load-flow problems in power stability by ϵ–coupling method. *Proc. IEE, Part D.*, **122**, 801–805.

99. Meirovich, L., (1967). *Analytical Methods in Vibrations*, Macmillan, New York, 1967.

100. Meirovich, L. and H. Baruh, (1983). On the problem of observation spillover in self-adjoint distributed parameter systems. *J. Optimization Theory and Applications*, **39**, 269–291.

101. Mil'shtein, G. (1964). Successive approximation for solution of one optimum problem, *Auto. and Rem. Control*, **25**, 298–306.

102. Mohler, R. (1970). Natural bilinear control processes. *IEEE Trans. Syst. Sci. Cybern.* **SSC-6**, 192–197.

103. Mohler, R. and C. Chen, *Optimal Control of Nuclear Reactors*, Academic Press, New York, 1970.

104. Mohler, R. (1973). *Bilinear Control Processes*, Academic Press, New York.

105. Mohler, R., (1974). Biological modeling with variable compartmental structure. *IEEE Trans. Automatic Control*, **AC-19**, 922–926.

106. Mohler, R. (1991). *Nonlinear Systems — Applications to Bilinear Control*. Prentice-Hall, Englewood Cliffs.

107. Mohler, R. and W. Koludziej, (1980). An overview of bilinear systems theory and applications. *IEEE Trans. Systems, Man and Cybernetics,* **SMC-10**, 683–688.

108. Mohler, R. and W. Koludziej, (1981). Optimal control of a class of nonlinear stochastic systems. *IEEE Trans. Automatic Control,* **AC-26**, 1048–1054.

109. **Ohta**, Y. and D. Siljak, (1985). Overlapping block diagonal dominance and existence of Lyapunov functions. *J. Math. Anal. Appl.,* **112**, 396–410.

110. O'Malley, R., (1974a). Boundary layer methods for certain nonlinear singularly perturbed optimal control problems. *J. Math. Anal. Appl.,* **45**, 468–484.

111. O'Malley, R., (1974b). *Introduction to Singular Perturbation,* Academic Press, New York.

112. Ozguner, U. and W. Perkins, (1979). A series solution to the Nash strategies for large scale interconnected systems. *Automatica,* **13**, 313–315.

113. Ozguner, U., (1979). Near-optimal control of composite systems: the multi time-scale approach. *IEEE Trans. Automatic Control,* **AC-24**, 652–655.

114. **Petkovski**, D., and M. Rakic, (1979). A series solution of feedback gains for output constrained regulators. *Int. J. Control,* **29**, 661–669.

115. Petrovic, B., and Z. Gajic, (1988). Recursive solution of linear-quadratic Nash games for weakly interconnected systems. *J. Optimization Theory and Applications,* **56**, 463–477.

116. **Quin**, J., (1980). Stabilization of bilinear systems by quadratic feedback controls. *J. Math. Anal. Appl.,* **75**, 66–80.

117. Qureshi, M. and Z. Gajic, (1991). Boundary value problem of linear weakly coupled systems. *Proc. Allerton Conference on Communication, Control and Computing,* 455–462, Urbana, IL.

118. Qureshi, M., X. Shen, and Z. Gajic, (1991). Open-loop control of singularly perturbed discrete systems. *Proc. Conf. on Information Sciences and Systems,* 151–155, Baltimore, MD.

119. Qureshi, M., (1992). *Parallel Algorithms for Discrete Singularly Perturbed and Weakly Coupled Filtering and Control Problems.* Ph. D. Dissertation, Rutgers University, 1992.

120. Qureshi, M. and Z. Gajic, (1992). A new version of the Chang transformation. *IEEE Trans. Automatic Control,* **AC-37,** 800–801.

121. **Ran,** A. and L. Rodman, (1988). On parameter dependence of solutions of algebraic Riccati equations. *Mathematics for Control, Signals and Systems,* **1,** 269–284.

122. Rutkowski, J., (1995). *Newton Method for Solving Algebraic Riccati Equations of Small Parameter Control Systems,* M.S. Thesis, Rutgers University.

123. Rutkowski, J. and Z. Gajic, (1993). Newton method for solving singularly perturbed algebraic Riccati equation. *Proc. IEEE Regional Conf. on Control Systems,* 196–199, Newark, NJ.

124. Ryan, E., (1984). Optimal feedback control of bilinear systems. *J. Optimization Theory and Applications,* **44,** 333–362.

125. Ryan, E. and N. Buckingham, (1983). On asymptotically stabilizing feedback control of bilinear systems. *IEEE Trans. Automatic Control,* **AC-28,** 863–864.

126. **Saberi,** A. and H. Khalil, (1984). Quadratic-type Lyapunov functions for singularly perturbed systems. *IEEE Trans. Automatic Control,* **AC-29,** 542–550.

127. Saberi, A. and H. Khalil, (1985). Stabilization and regulation of nonlinear singularly perturbed systems — composite control. *IEEE Trans. Automatic Control,* **AC-30,** 739–747.

128. Sage, A. and C. White, (1977). *Optimum Systems Control,* Prentice Hall, Englewood Cliffs, NJ.

129. Saksena, V. and J. Cruz, (1981a). A multimodel approach to stochastic Nash games. *Automatica,* **17,** 295–305.

130. Saksena, V. and J. Cruz, (1981b). Nash strategies in decentralized control of multiparameter singularly perturbed large scale systems. *Large Scale Systems,* **2,** 219–234.

131. Saksena, V. and T. Basar, (1982). A multimodel approach to stochastic team problems. *Automatica*, **18**, 713–720.

132. Schwarz, H., H. Dorissen, and L. Guo, (1988). Bilinearization of non-linear systems. in *Systems Analysis and Simulation*, 89–96, Akademie-Verlag, Berlin.

133. Sezer, M., and D. Siljak, (1986). Nested ϵ-decomposition and clustering of complex systems. *Automatica*, **22**, 321–331.

134. Sezer, M., and D. Siljak, (1991). Nested epsilon decomposition of linear systems: Weakly coupled and overlapping blocks. *SIAM J. Matrix Anal. Appl.*, **3**, 521–533.

135. Slemrod, M., (1978). Stabilization of bilinear control systems with application to nonconservative problems in elasticity. *SIAM J. Control and Optimization*, **16**, 131–141.

136. Skataric, D., Z. Gajic, and D. Petkovski, (1991). Reduced-order solution for a class of linear quadratic optimal control problems. *Proc. Allerton Conference on Communication, Control and Computing*, 440–447, Urbana, IL.

137. Skataric, D. and Z. Gajic, (1992). Linear control of nearly singularly perturbed hydro power plants. *Automatica*, **28**, 159–163.

138. Skataric, D., Z. Gajic, and D. Arnautovic, (1993). Reduced-order design of optimal controllers for quasi weakly coupled linear systems. *Control — Theory and Advanced Technology*, **9**, 481–490.

139. Shen, X. and Z. Gajic, (1990a). Near-optimum steady state regulators for stochastic linear weakly coupled systems. *Automatica*, **26**, 919–923.

140. Shen, X. and Z. Gajic, (1990b). Optimal reduced-order solution of the weakly coupled discrete Riccati equation. *IEEE Trans. on Automatic Control*, **AC-35**, 600–602.

141. Shen, X. and Z. Gajic, (1990c). Approximate parallel controllers for discrete weakly coupled linear stochastic systems. *Optimal Control Appl. & Methods*, **11**, 345–354.

142. Siljak, D., (1991). *Decentralized Control of Complex Systems*. Academic Press, Cambridge, MA.

143. Srikant, R. and T. Basar, (1989). Optimal solutions in weakly coupled multiple decision maker Markov chains. *Proc. Conference on Decision and Control*, 168–173, Tampa, FL.

144. Srikant, R. and T. Basar, (1991). Iterative computation of noncooperative equilibria in nonzero-sum differential games with weakly coupled players. *J. Optimization Theory and Applications*, **71**, 137–168.

145. Srikant, R. and T. Basar, (1992a). Asymptotic solutions of weakly coupled stochastic teams with nonclassical information. *IEEE Trans. Automatic Control*, **AC–37**, 163–173.

146. Srikant, R. and T. Basar, (1992b). Sequential decomposition and policy iteration schemes for M-player games with partial weak coupling. *Automatica*, **28**, 95–105.

147. Su, W. and Z. Gajic, (1991). Reduced-order solution to the finite time optimal control problems of linear weakly coupled systems. *IEEE Trans. Automatic Control*, **AC-36**, 498–501.

148. Su, W. and Z. Gajic, (1992). Decomposition method for solving weakly coupled algebraic Riccati equation. *AIAA J. Guidance, Dynamics and Control*, **15**, 536–538.

149. Su, W., Z. Gajic, and X. Shen, (1992a). The recursive reduced-order solution of an open-loop control problem of linear singularly perturbed systems. *IEEE Trans. Automatic Control*, **AC-37**, 279–281.

150. Su, W., Z. Gajic, and X. Shen, (1992b). The exact slow-fast decomposition of the algebraic Riccati equation of singularly perturbed systems. *IEEE Trans. Automatic Control*, **AC-37**, 1456–1459.

151. Sundareshan, M. and R. Fundkowski, (1985). Periodic optimization of a class of bilinear systems with application to control of cell proliferation and cancer therapy. *IEEE Trans. Systems Man and Cybernetics*, **SMC-15**, 102–115.

152. Sundareshan, M. and R. Fundkowski, (1986). Stability and control of a class of compartmental systems with application to cell proliferation and cancer therapy. *IEEE Trans. Automatic Control*, **AC-31**, 1022–1031.

153. Sussmann, H., (1976). Semigroup representation, bilinear approximation of input-output maps and generalized input. in *Mathematical System Theory*, G. Marchesini and S. Mitter, Eds., Springer Verlag, New York, 1976.

154. Suzuki, M., (1981). Composite control for singularly perturbed systems. *IEEE Trans. Automatic Control*, **AC-26**, 505–507.

155. Swamy, K. and T. Tran, (1979). Deterministic and stochastic control of discrete time bilinear systems. *Automatica*, **15**, 677–682.

156. **Tzafestas**, S. and K. Anagnostou, (1984a). Stabilization of singularly perturbed strictly bilinear systems. *IEEE Trans. Automatic Control*, **AC-29**, 943–946.

157. Tzafestas, S. and K. Anagnostou, (1984b). Stabilization of ϵ-coupled bilinear systems using state feedback control. *Int. J. Systems Science*, **15**, 639–646.

158. Tzafestas, S., K. Anagnostou, and T. Pimenides, (1984). Stabilizing optimal control of bilinear systems with generalized cost. *Optimal Control Applications & Methods*, **5**, 111–117.

159. **Yaz**, E., (1992). Full and reduced-order observer design for discrete stochastic bilinear systems. *IEEE Trans. Automatic Control*, **AC-37**, 503–505.

160. Yi, G., Y. Hwang, H. Chang, and K. Lee, (1989). Computer control of cell mass concentration in continuous culture. *Automatica*, **25**, 243–249.

161. Ying, Y., M. Rao, and X. Shen, (1992). Bilinear decoupling control and its industrial application. *Proc. American Control Conference*, 1163–1167, Chicago, IL.

162. Ying, Y., M. Rao, and Y. Sun, (1993). Suboptimal control for bilinear systems. *Optimal Control Applications & Methods*, **1**, 195–202.

163. **Vaisbord**, E., (1963). An approximate method for the synthesis of optimal control, *Auto. and Rem. Control*, **24**, 1626–1632.

164. **Washburn**, H. and J. Mendel, (1980). Multistage estimation of dynamical and weakly coupled systems in continuous-time linear systems. *IEEE Trans. Automatic Control*, **AC-25**, 71–76.

165. Wei, K. and A. Pearson, (1978). On minimum energy control of communicative bilinear systems. *IEEE Trans. Automatic Control*, **AC-23**, 1020–1023.

166. Wiener, N. (1948). *Cybernetics*, MIT Press, Cambridge.

167. Wilde, R. and P. Kokotovic, (1973). Optimal open-loop and closed-loop control of singularly perturbed linear systems. *IEEE Trans. Automatic Control*, **AC-17**, 616–625.

168. Williamson, D., (1977). Observation of bilinear systems with application to biological control. *Automatica*, **13**, 243–254.

169. Wonham, W., (1967). Optimal stationary control of linear systems with state-dependent noise. *SIAM J. Control*, **5**, 486–500.

170. Wonham, W., (1968). On a matrix Riccati equation of stochastic control. *SIAM J. Control*, **6**, 681–697.

171. **Zhuang**, J. and Z. Gajic, (1991). Stochastic multimodel strategy with perfect measurements. *Control — Theory and Advanced Technology*, **7**, 173–182.

Index

Lecture Notes in Control and Information Sciences

Edited by M. Thoma

1992–1995 Published Titles:

Vol. 180: Kall, P. (Ed.)
System Modelling and Optimization.
Proceedings of the 15th IFIP Conference,
Zurich, Switzerland, September 2-6, 1991
969 pp. 1992 [3-540-55577-3]

Vol. 181: Drane, C.R.
Positioning Systems - A Unified Approach
168 pp. 1992 [3-540-55850-0]

Vol. 182: Hagenauer, J. (Ed.)
Advanced Methods for Satellite and Deep
Space Communications. Proceedings of
an International Seminar Organized by
Deutsche Forschungsanstalt für Luft-und
Raumfahrt (DLR), Bonn, Germany,
September 1992
196 pp. 1992 [3-540-55851-9]

Vol. 183: Hosoe, S. (Ed.)
Robust Control. Proceesings of a Workshop
held in Tokyo, Japan, June 23-24, 1991
225 pp. 1992 [3-540-55961-2]

Vol. 184: Duncan, T.E.; Pasik-Duncan, B.
(Eds)
Stochastic Theory and Adaptive Control.
Proceedings of a Workshop held in
Lawrence, Kansas, September 26-28,
1991
500 pp. 1992 [3-540-55962-0]

Vol. 185: Curtain, R.F. (Ed.); Bensoussan,
A.; Lions, J.L.(Honorary Eds)
Analysis and Optimization of Systems:
State and Frequency Domain Approaches
for Infinite-Dimensional Systems.
Proceedings of the 10th International
Conference, Sophia-Antipolis, France, June
9-12, 1992.
648 pp. 1993 [3-540-56155-2]

Vol. 186: Sreenath, N.
Systems Representation of Global Climate
Change Models. Foundation for a Systems
Science Approach.
288 pp. 1993 [3-540-19824-5]

Vol. 187: Morecki, A.; Bianchi, G.;
Jaworeck, K. (Eds)
RoMansSy 9: Proceedings of the Ninth
CISM-IFToMM Symposium on Theory and
Practice of Robots and Manipulators.
476 pp. 1993 [3-540-19834-2]

Vol. 188: Naidu, D. Subbaram
Aeroassisted Orbital Transfer: Guidance
and Control Strategies
192 pp. 1993 [3-540-19819-9]

Vol. 189: Ilchmann, A.
Non-Identifier-Based High-Gain Adaptive
Control
220 pp. 1993 [3-540-19845-8]

Vol. 190: Chatila, R.; Hirzinger, G. (Eds)
Experimental Robotics II: The 2nd
International Symposium, Toulouse,
France, June 25-27 1991
580 pp. 1993 [3-540-19851-2]

Vol. 191: Blondel, V.
Simultaneous Stabilization of Linear
Systems
212 pp. 1993 [3-540-19862-8]

Vol. 192: Smith, R.S.; Dahleh, M. (Eds)
The Modeling of Uncertainty in Control
Systems
412 pp. 1993 [3-540-19870-9]

Vol. 193: Zinober, A.S.I. (Ed.)
Variable Structure and Lyapunov Control
428 pp. 1993 [3-540-19869-5]

Vol. 194: Cao, Xi-Ren
Realization Probabilities: The Dynamics of
Queuing Systems
336 pp. 1993 [3-540-19872-5]

Vol. 195: Liu, D.; Michel, A.N.
Dynamical Systems with Saturation
Nonlinearities: Analysis and Design
212 pp. 1994 [3-540-19888-1]

Vol. 196: Battilotti, S.
Noninteracting Control with Stability for
Nonlinear Systems
196 pp. 1994 [3-540-19891-1]